KNOWLEDGE ENCYCLOPEDIA
SCIENCE!

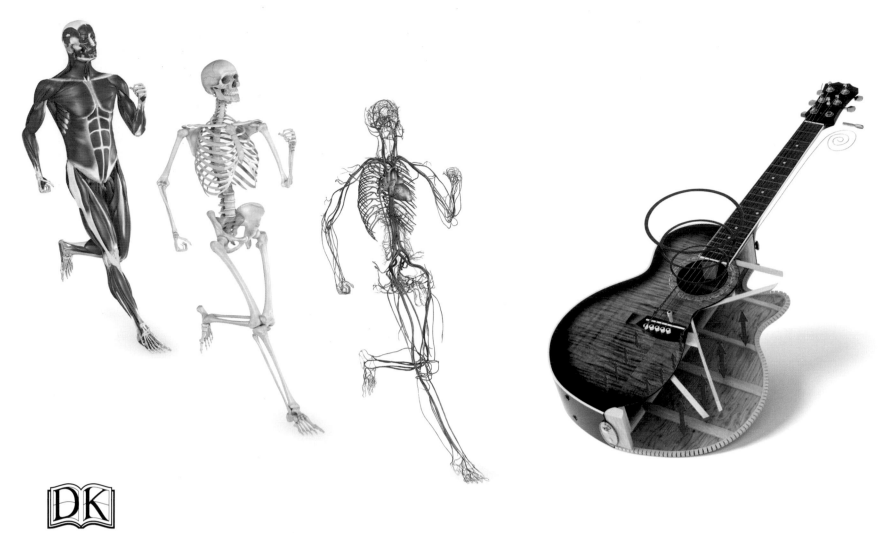

DK

KNOWLEDGE ENCYCLOPEDIA
SCIENCE!

Written by: Abigail Beall, Jack Challoner, Adrian Dingle, Derek Harvey, Bea Perks
Consultant: Jack Challoner

Illustrators: Peter Bull, Jason Harding, Stuart Jackson-Carter – SJC Illustration,
Jon @ KJA, Arran Lewis, Sofian Moumene, Alex Pang, Jack Williams

DK UK:

Senior Editor Georgina Palffy

Senior Art Editor Stefan Podhorodecki

Editors Vicky Richards, Anna Streiffert Limerick, Alison Sturgeon

Designers David Ball, Gregory McCarthy, Sadie Thomas

Managing Editor Francesca Baines

Managing Art Editor Philip Letsu

Jacket Design Development Manager Sophia MTT

Jacket Editor Amelia Collins

Jacket Designer Surabhi Wadhwa Gandhi

Producer (Pre-Production) Jacqueline Street

Producer Jude Crozier

Publisher Andrew Macintyre

Art Director Karen Self

Associate Publishing Director Liz Wheeler

Design Director Philip Ormerod

Publishing Director Jonathan Metcalf

DK India:

Managing Jackets Editor Saloni Singh

Jacket Designer Tanya Mehrotra

Senior DTP Designer Harish Aggarwal

Jackets Editorial Coordinator Priyanka Sharma

Picture Research Manager Taiyaba Khatoon

Picture Researcher Deepak Negi

First published in Great Britain in 2018
by Dorling Kindersley Limited,
80 Strand, London WC2R 0RL

Copyright © 2018 Dorling Kindersley Limited

A Penguin Random House Company

4 6 8 10 9 7 5

013 – 308119 – August/2018

ISBN: 978-0-2413-1781-5

Printed and bound in China

A WORLD OF IDEAS:
SEE ALL THERE IS TO KNOW

www.dk.com

CONTENTS

MATTER

ENERGY & FORCES

LIFE

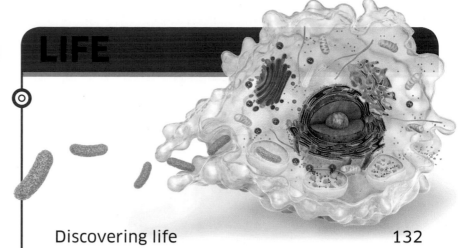

REFERENCE

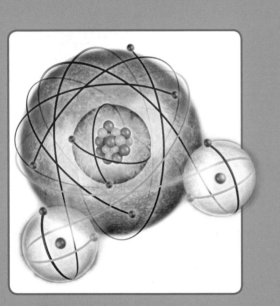

MATTER

The ground beneath your feet, the air around you, and the stars in the sky are made of matter. You are made of matter, too. All matter is made of minute particles called atoms, which join together in countless ways to form an astonishing variety of substances.

1909

pH scale invented
Danish chemist Søren Peder Lauritz Sørensen invents the pH scale, which is used to judge whether a substance is an acid, neutral, or base.

1898

New elements
Polish-French scientist Marie Curie and her husband Pierre discover two new radioactive elements, radium and polonium. Radium is later used in radiotherapy to treat cancer.

1913

Electron shells
The Danish scientist Niels Bohr proposes a model of the atom that shows how electrons occupy shells and orbit around the nucleus.

MODERN TIMES

Modern chemistry
Advances in technology allowed chemists and other scientists to invent new materials by reproducing natural materials synthetically or rearranging atoms through nanotechnology.

CATHODE RAY TUBE

1897

Discovery of electrons
English scientist JJ Thomson discovers electrons using a cathode ray tube. This is the first step towards understanding the structure of atoms.

The Atomic Age
The discovery of radioactivity led to a better understanding of what lies inside an atom, and more research into subatomic particles. This knowledge was put to use in medicine and healthcare.

1890 – 1945

Discovering matter

Thousands of years of questioning, experimentation, and research have led to our understanding of matter as we know it today.

Following the earliest explorations of matter by our prehistoric ancestors, Greek philosophers were among the first people to attempt to classify matter and explain its behaviour. Over time, scientists found more sophisticated ways of analysing different types of matter and discovered many of the elements. The Industrial Revolution saw the invention of new synthetic materials using these elements, while greater understanding of the structure of atoms led to significant advances in medicine. New substances and materials with particularly useful properties are still being discovered and invented to this day.

1772 / 1774

Discovery of oxygen
Swedish chemist Carl Scheele builds a contraption to capture oxygen by heating various compounds together. English scientist Joseph Priestley also discovers oxygen by showing that a candle can't burn without it.

1789

Antoine Lavoisier
French chemist Antoine Lavoisier publishes *Elements of Chemistry*, which lists the 33 known elements divided into four types: gases, metals, non-metals, and earths.

SCHEELE'S OXYGEN APPARATUS

Timeline of discoveries
From prehistory to the present day, people have sought to understand how matter behaves and to classify different types. Over the years, this has led to the discovery of new matter and materials.

Prehistory to antiquity
The earliest discoveries of how matter behaves were made not by scientists, but by our prehistoric ancestors trying to survive. During antiquity, philosophers spent a lot of time trying to work out what matter is.

Making fire
Our ancestors learn to make fire using combustion (although they don't know that at the time).

Copper and bronze
Smelting of copper (extracting it from its ore through heat) is discovered. Bronze (copper smelted with tin) is first produced in 3200 BCE.

Greek philosophers
Empedocles suggests that everything is made of four elements: air, earth, fire, and water. Democritus suggests that all matter consists of atoms.

BEFORE 500 CE

790,000 BCE   **3200 BCE**   **420 BCE**

1661 Robert Boyle's *The Sceptical Chymist* develops a **theory of atoms**.

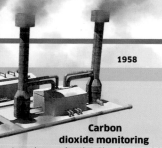

1958

1985

2004

BUCKYBALL

GRAPHENE

Carbon dioxide monitoring
American scientist Charles David Keeling starts to monitor the rise of carbon dioxide in the atmosphere. His Keeling Curve graph is still used to study climate change.

Buckyball discovery
Scientists at Rice University, Houston, USA, discover a new form of carbon called buckminsterfullerene, or buckyball.

World's thinnest material
Graphene (a layer of carbon atoms just one atom thick) is produced at the University of Manchester, UK. It is the world's thinnest material, but 200 times stronger than steel.

1945 – PRESENT

1870

1869

Synthetic materials
The first synthetic materials made from cellulose are invented: celluloid (mouldable plastic) in 1870 and viscose rayon in 1890.

Mendeleev's periodic table
Russian chemist Dmitri Mendeleev arranges the 59 known elements into groups based on their atomic mass and properties. This periodic table enables him to predict the discovery of three more elements.

VISCOSE, SYNTHETIC SILK

GAY-LUSSAC EXPERIMENTING WITH AIR PRESSURE IN HOT-AIR BALLOON

1890-1945

1803

Industrial Revolution
Driven by the thirst for modernization, chemists identified more elements and invented ways to use them in medicine, in creating new materials, and in advanced industrial technologies.

Dalton's atomic theory
English chemist John Dalton argues that all matter is composed of atoms and atoms of the same element are identical. He compiles a list of elements based on their atomic mass, then known as atomic weight.

Structure of water
French chemist Joseph Louis Gay-Lussac experiments with gases and pressure and finds that water is made up of two parts hydrogen and one part oxygen.

1805

DALTON'S ATOMIC MODELS

19TH CENTURY

1800 – 1890

1527

Age of Discovery
The Renaissance brought both rediscovery of antique knowledge and a quest for fresh ideas. Scientists began to test, experiment, and document their ideas, publishing their findings and working hard to classify matter.

Salts, sulfurs, and mercuries
Swiss chemist Theophrastus von Hohenheim works out a new classification for chemicals, based on salts, sulfurs, and mercuries.

17TH CENTURY

1600 – 1800

Classifying elements
Persian physician Al-Razi divides elements into spirits, metals, and minerals depending on how they react with heat.

Middle Ages
In Asia and the Islamic world, alchemists experimented to find the elixir of life and to make gold. By the late Middle Ages, European alchemists were working towards the same goal.

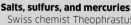

Gunpowder
While they are looking for the elixir of life, Chinese alchemists accidentally invent gunpowder by mixing saltpetre with sulfur and charcoal.

MIDDLE AGES

500 CE – 1600

855 CE

900

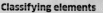

WHAT IS MATTER?

The air around you, the water you drink, the food you eat, your own body, the stars and the planets – all of these things are matter. There is clearly a huge variety of different types of matter, but it is all made of tiny particles called atoms, far too small to see. About ninety different kinds of atom join together in many combinations to make all the matter in the Universe.

PARTICLES OF MATTER

Matter is made of atoms – but in many substances, those atoms are combined in groups called molecules, and in some they exist as ions: atoms that carry an electric charge. Both atoms and ions can bond together to form compounds.

Atoms and molecules

An atom is incredibly small: you would need a line of 100,000 of them to cover the width of a human hair. Tiny though they are, atoms are made of even smaller particles: protons, neutrons, and electrons. Different kinds of atom have different numbers of these particles. Atoms often join, or bond, in groups called molecules. A molecule can contain atoms of the same kind or of different kinds.

It's a matter of water

Water is one of the most abundant substances on Earth. More than two thirds of Earth's surface is covered by water. Animals contain lots of water, too – nearly two thirds of a cat's mass is water, for example. Water is made up of H_2O molecules, each made up of atoms of oxygen and hydrogen.

WATER MOLECULE (H_2O)

Oxygen atom (O)

Hydrogen atom (H)

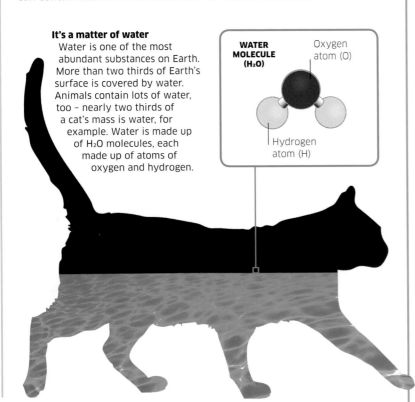

ELEMENTS, COMPOUNDS, AND MIXTURES

Everything around us is matter, but it is a bit more complex than that. Elements can exist on their own, but usually bond together chemically with other elements to form compounds, or appear in mixtures (substances in which the "ingredients" are not chemically bonded, but simply mixed together). A mixture can consist of two or more elements, an element and a compound, or two or more separate compounds.

What's what?

Everything can be sorted into different categories of matter, depending on whether it is a pure substance or a mixture of different substances. This diagram shows the main types.

Cut diamond

Pure substances
Matter is pure if it is made of just one kind of substance. That substance can be an element or a compound. Diamond, a form of the element carbon, is a pure substance. So is salt (sodium chloride), a compound of the elements sodium and chlorine.

Elements
An element, such as gold, is a pure substance, made of only one kind of atom. Iron, aluminium, oxygen, carbon, and chlorine are other examples of elements. All elements have different properties, and are sorted into a chart called the periodic table (see pp.28-29).

Compounds
A compound is a pure substance that consists of atoms of different elements bonded together. In any particular compound, the ratio of the different kinds of atoms is always the same. In salt there are equal numbers of sodium and chlorine atoms (1:1), while water contains twice as many hydrogen as oxygen atoms (2:1).

Stainless steel, an alloy of iron, carbon, and chromium, is a homogeneous mixture.

Homogeneous mixtures
In a homogeneous mixture, particles of different substances are mixed evenly, so the mixture has the same composition throughout. They can be solid (steel), liquid (honey), or gas (air).

Solutions
All homogeneous mixtures are solutions, but the most familiar are those where a solid has been dissolved in a liquid. An example is salty water – in which the salt breaks down into ions that mix evenly among the water molecules. In sugary drinks the sugar is also dissolved – no grains of sugar float about in the solution.

The air in a balloon is a homogeneous mixture of several gases, mostly the elements nitrogen and oxygen.

Matter
Matter can be solid, liquid, or gas. Most of the matter around you, from planets to animals, is composed of mixtures of different substances. Only a few substances exist naturally in completely pure form.

A frog is made of compounds and mixtures.

An ice cream is an impure substance – a mixture of many different ingredients.

A sandwich is a mixture of several substances.

Impure substances
If a substance is impure, it means that something has got mixed into it. For example, pure water consists of only hydrogen and oxygen. But tap water contains minerals, too, which makes it an impure substance. All mixtures are impure substances.

Mixtures
There are many different kinds of mixtures, depending on what substances are in the mix and how evenly they mix. The substances in a mixture are not bound together chemically, and can be separated. Rocks are solid mixtures of different minerals that have been pressed or heated together.

A leaf is a very complex uneven mixture.

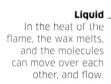

Muddy water is a suspension: it may look evenly mixed at first, but the larger mud particles soon separate out.

Heterogeneous mixtures
In a heterogeneous mixture, particles of different substances are mixed unevenly. Examples are concrete (a mixture of sand, cement, and stone) and sand on a beach, which consists of tiny odd-sized particles of eroded rock, sea shells, and glass fragments.

Suspensions
Suspensions are liquids that contain small particles that do not dissolve. If they are shaken, they can appear evenly mixed for a short time, but then the particles separate out and you can see them with your naked eye.

Colloids
A colloid looks like an even mixture, but no particles have been completely dissolved. Milk, for example, consists of water and fat. The fat does not dissolve in water, but floats about in minute blobs that you cannot see without a microscope. A cloud is a colloid of tiny water droplets mixed in air.

◉ STATES OF MATTER
Most substances exist as solids, liquids, or gases – or as mixtures of these three states of matter. The particles of which they are made (the atoms, molecules, or ions) are in constant motion. The particles of a solid vibrate but are held in place – that's why a solid is rigid and keeps its shape. In a liquid, the particles are still attracted to each other, but can move over each other, making it fluid. In a gas, the particles have broken free from each other, and move around at high speed.

Changing states of matter
With changes in temperature, and sometimes in pressure, one state can change into another. If it is warm, a solid ice cube melts into liquid water. If you boil the water, it turns into gaseous steam. When steam cools down, it turns back into a liquid, such as the tiny droplets of mist forming on a bathroom window. Only some substances, including candle wax, exist in all three states.

Gases
Near the wick, the temperature is high enough to vaporize the liquid wax, forming a gas of molecules that can react with the air. This keeps the flame burning.

Liquid
In the heat of the flame, the wax melts, and the molecules can move over each other, and flow.

Solid
Solid wax is made of molecules held together. Each wax molecule is made of carbon and hydrogen atoms.

Plasma, the fourth state of matter
When gas heats up to a very high temperature, electrons break free from their atoms. The gas is now a mixture of positively charged ions and negatively charged electrons: a plasma. A lightning bolt is a tube of plasma because of the extremely high temperature inside it. In space, most of the gas that makes up the Sun, and other stars in our Universe, is so hot it is plasma.

Atomic proportions

You would have to enlarge an atom to a trillion times its size to make it as big as a football stadium. Even at that scale, the atom's electrons would be specks of dust flying around the stadium, and its nucleus would be the size of a marble.

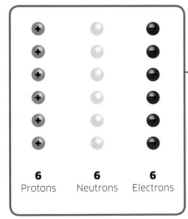

Size of the nucleus if the atom was the size of a stadium.

Atoms

You, and all the things around you, are made of tiny particles called atoms – particles so minuscule that even a small grain of sand is made up of trillions of them.

Atoms were once thought to be the smallest possible parts of matter, impossible to split into anything smaller. But they are actually made of even smaller particles, called protons, neutrons, and electrons. Atoms join, or bond, in many different ways to make every different kind of material. A pure substance, consisting of only one type of atom, is called an element. Some familiar elements include gold, iron, carbon, neon, and oxygen. To find out more about the elements, see pages 28–41.

Atomic structure

The nucleus at the centre of an atom is made of protons and neutrons. The protons carry a positive electric charge. The neutrons carry no charge – they are neutral. Around the nucleus are the electrons, which carry a negative electric charge. It is the force between the positively charged protons and the negatively charged electrons that holds an atom together.

Particles of an atom

Every atom of an element has the same number of electrons as it has protons, but the number of neutrons can be different. Below are the particles of one atom of the element carbon.

6	**6**	**6**
Protons	Neutrons	Electrons

Carbon atom

The number of protons in an atom's nucleus is called the atomic number. This defines what an element is like: each element has a different atomic number, as shown in the periodic table (see pp.28-29). For the element carbon, shown here, the atomic number is 6. An atom's number of electrons is also equal to its atomic number.

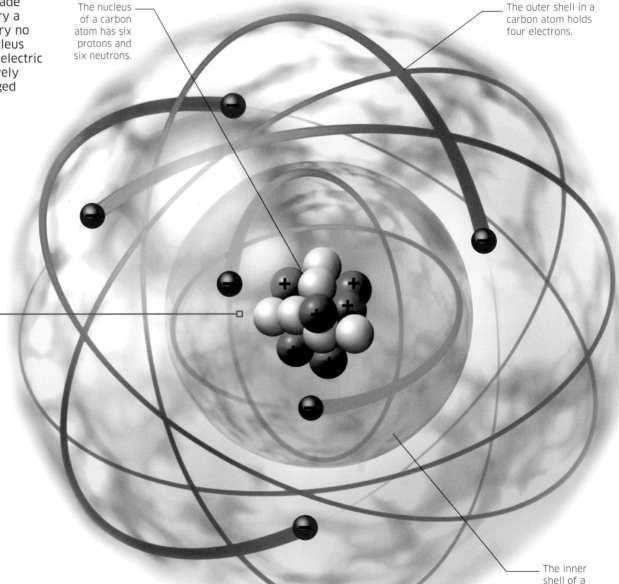

The nucleus of a carbon atom has six protons and six neutrons.

The outer shell in a carbon atom holds four electrons.

The inner shell of a carbon atom holds two electrons.

Atoms of the **element helium**
are **the smallest** of all **atoms**.

1803 The year schoolteacher **John Dalton** presented his
theory about **what atoms are** and what they do.

13

Electrons and electron shells

An atom's electrons are arranged around the nucleus in shells. Each shell can hold a certain number of electrons before it is full: the inner shell can hold two, the next shell eight, the third one 18, and so on. The heaviest atoms, with large numbers of electrons, have seven shells. Atoms that don't have full outer shells are unstable. They seek to share, or exchange, electrons with other atoms to form chemical compounds. This process is known as a chemical reaction. Atoms with a filled outer shell are stable, and therefore very unreactive.

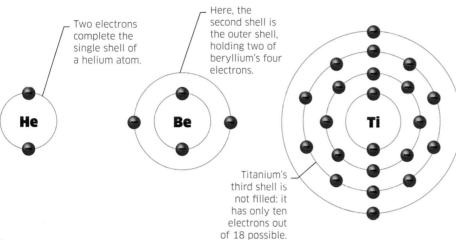

Two electrons complete the single shell of a helium atom.

Here, the second shell is the outer shell, holding two of beryllium's four electrons.

Titanium's third shell is not filled: it has only ten electrons out of 18 possible.

Helium
The gas helium has the atomic number 2. All its atoms have two electrons, which is the maximum number the first shell can hold. With a full outer shell, helium atoms are very unreactive.

Beryllium
The second shell of an atom can hold up to eight electrons. The metal beryllium (atomic number: 4) has a filled inner shell, but only two electrons in its outer shell, making it quite reactive.

Titanium
The metal titanium (atomic number: 22) has four shells. It has two electrons in its outer shell, even though the third shell is not full. It is quite common for metals to have unfilled inner shells.

Atomic mass and isotopes

The mass of an atom is worked out by counting the particles of which it is made. Protons and neutrons are more than 1,800 times heavier than electrons, so scientists only take into account those heavier particles, and not the electrons. All atoms of a particular element have the same number of protons, but there are different versions of the atoms, called isotopes, that have different numbers of neutrons. The relative atomic mass of an element is the average of the different masses of all its atoms.

Isotopes of sodium
All atoms of the element sodium (atomic number 11) have 11 protons, and nearly all have 12 neutrons. So the relative atomic mass is very close to 23, but not exactly.

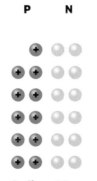

Sodium-22
The sodium isotope with 11 neutrons in its atoms has a mass of 22.

Sodium-23
Sodium-23, the most common sodium isotope, has 11 protons and 12 neutrons.

Sodium-24
This sodium isotope has a mass of 24: 11 protons and 13 neutrons.

Atoms and matter

It is difficult to imagine how atoms make the world around you. Everyday objects don't look as if they consist of tiny round bits joined together: they look continuous. It can help to zoom in closer and closer to an everyday material, such as paper, to get the idea.

Paper
Paper is made almost entirely of a material called cellulose, which is produced inside plant cells, usually from trees. Cellulose is hard-wearing and can absorb inks and paints.

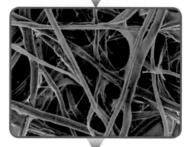

Cellulose fibre
Cellulose forms tiny fibres, each about one thousandth of a millimetre in diameter. The fibres join together, making paper strong and flexible.

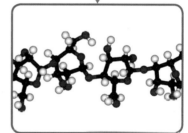

Cellulose molecule
Each cellulose fibre is made of thousands of molecules. A cellulose molecule is a few millionths of a millimetre wide. It is made of atoms of different elements: carbon (black), oxygen (red), and hydrogen (white).

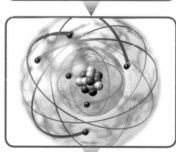

Carbon atom
A typical cellulose molecule contains a few thousand carbon atoms. Each carbon atom has six electrons that form bonds with atoms of the other two elements.

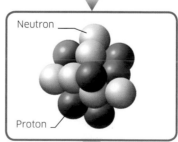

Neutron

Proton

Nucleus
Most of the carbon atom is empty space. Right at the centre, about one trillionth of a millimetre across, is the nucleus, made of six protons and six neutrons.

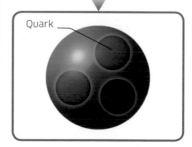

Quark

Quarks
Each particle in the nucleus is made of even smaller particles, called quarks. Each proton, and each neutron, is made of three quarks, held together by particles called gluons.

Molecules

A molecule consists of two or more atoms joined, or bonded, together. Many familiar substances, such as sugar or water, are made up of molecules. Molecules are so small that even a small drop of water contains trillions of them.

All the molecules of a particular compound (chemically bonded substance) are identical. Each one has the same number of atoms, from at least two elements (see pp.28–29), combined in the same way. The bonds that hold molecules together form during chemical reactions but they can be broken, as atoms react with other atoms and rearrange to form new molecules. It is not only compounds that can exist as molecules. Many elements exist as molecules, too, but then all the atoms that make up the molecule are identical, such as the pair of oxygen atoms that make up pure oxygen (O_2).

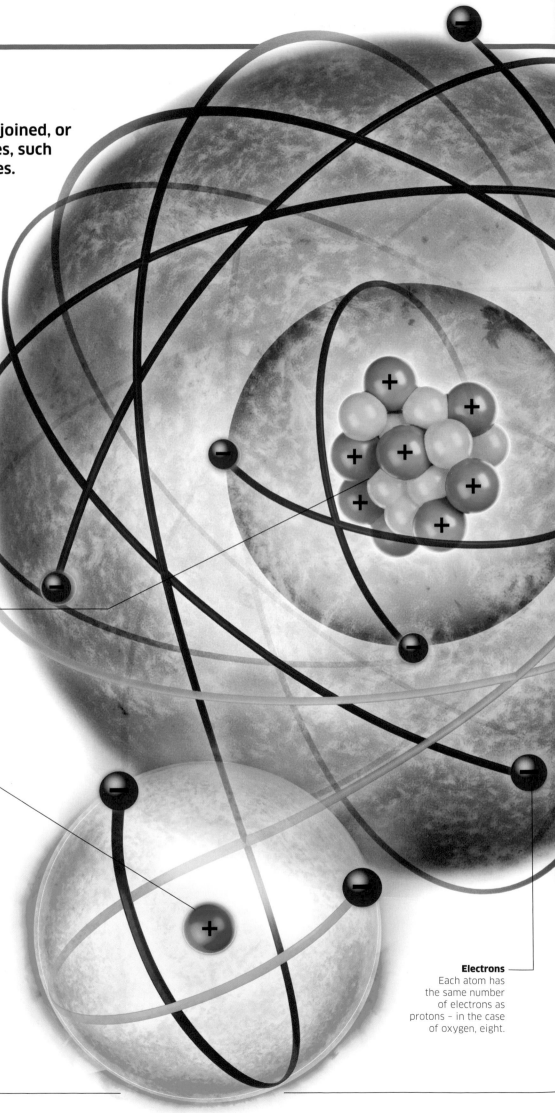

Nucleus of oxygen atom
The oxygen atom has eight protons and eight neutrons in its nucleus. Protons (shown in green) have a positive charge while neutrons (white) are neutral.

Nucleus of hydrogen atom
The hydrogen atom is the only atom that consists of just one proton in its nucleus, and does not contain any neutrons.

Electrons
Each atom has the same number of electrons as protons – in the case of oxygen, eight.

Water molecule

Imagine dividing a drop of water in half, and then in half again. If you could keep on doing this, you would eventually end up with the smallest amount of water: a water molecule. Every water molecule is made up of one oxygen atom and two hydrogen atoms. The atoms are held together as a molecule because they share electrons, in a type of chemical bond called a covalent bond (see also p.16).

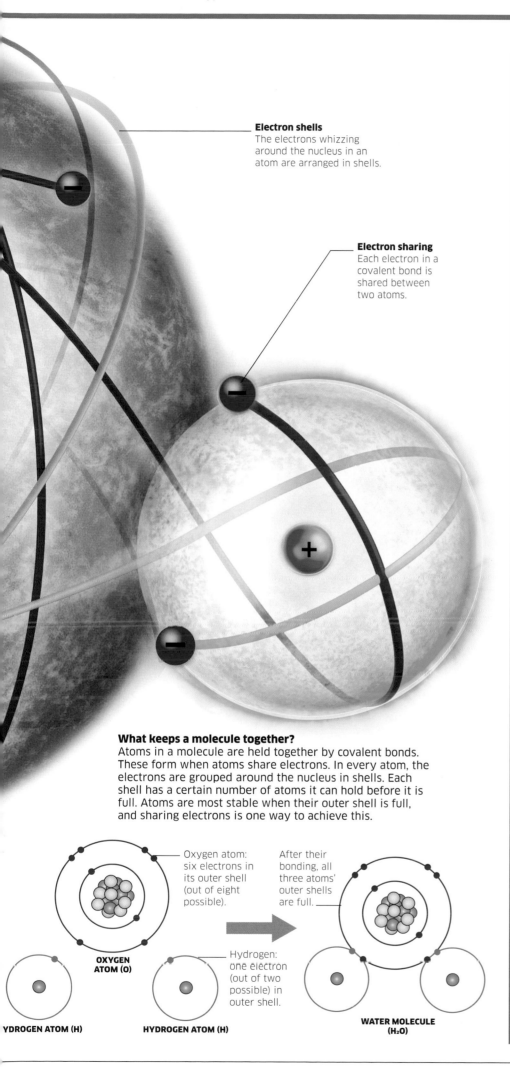

Electron shells
The electrons whizzing around the nucleus in an atom are arranged in shells.

Electron sharing
Each electron in a covalent bond is shared between two atoms.

What keeps a molecule together?

Atoms in a molecule are held together by covalent bonds. These form when atoms share electrons. In every atom, the electrons are grouped around the nucleus in shells. Each shell has a certain number of atoms it can hold before it is full. Atoms are most stable when their outer shell is full, and sharing electrons is one way to achieve this.

Oxygen atom: six electrons in its outer shell (out of eight possible).

After their bonding, all three atoms' outer shells are full.

OXYGEN ATOM (O)

Hydrogen: one electron (out of two possible) in outer shell.

YDROGEN ATOM (H)   **HYDROGEN ATOM (H)**

WATER MOLECULE (H₂O)

Elements and compounds

Most elements are made up of single atoms, but some are made of molecules of two or more identical atoms. When two elements react, their molecules form a new compound.

Oxygen
The gas oxygen (O₂) is made of molecules, each containing two oxygen atoms.

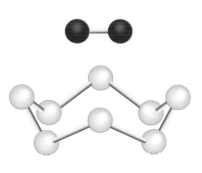

Sulfur
Pure sulfur (S), a solid, normally exists as molecules of eight sulfur atoms bonded together.

Sulfur dioxide (SO₂)
When sulfur and oxygen molecules react, their bonds break to make new bonds, and a new substance forms.

Representing molecules

Scientists have different ways of representing molecules to understand how chemical reactions happen. Here, a molecule of the gas compound methane (CH₄), made of one carbon atom and four hydrogen atoms, is shown in three ways.

Lewis structure
The simplest way to represent a molecule is to use the chemical symbols (letters), and lines for covalent bonds.

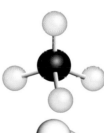

Ball and stick
Showing the atoms as balls and the bonds as sticks gives a three-dimensional representation of a molecule.

Space filling
This method is used when the space and shape of merged atoms in a molecule are more important to show than bonds.

Macromolecules

While some compounds are made of small molecules consisting of just a few atoms, there are many compounds whose molecules are made of thousands of atoms. This molecular model shows a single molecule of a protein found in blood, called albumin. It contains atoms of many different elements, including oxygen, carbon, hydrogen, nitrogen, and sulfur.

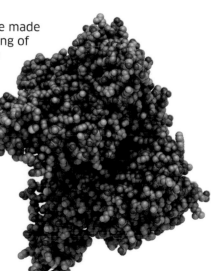

Bonding

Matter is made of atoms. Most of them are joined, or bonded, together. The bonds that hold atoms together are formed by the outermost parts of each atom: the electrons in the atom's outer shell.

There are three main types of bonding: ionic, covalent, and metallic. An ionic bond forms when electrons from one atom transfer to another, so that the atoms become electrically charged and stick together. A covalent bond forms when electrons are shared between two or more atoms. In a metal, the electrons are shared freely between many metal atoms. All chemical reactions involve bonds breaking and forming.

To bond or not to bond

The number of electrons an atom has depends upon how many protons are in its nucleus. This number is different for each element (see p.28). The electrons are arranged in "shells", and it is the electrons in the outermost shell that take part in bonding. An atom is stable when the outermost shell is full (see p.13). The atoms of some elements have outermost shells that are already full – they do not form bonds easily. But most atoms can easily lose or gain electrons, or share them with other atoms, to attain a full outer shell. These atoms do form bonds, and take part in chemical reactions.

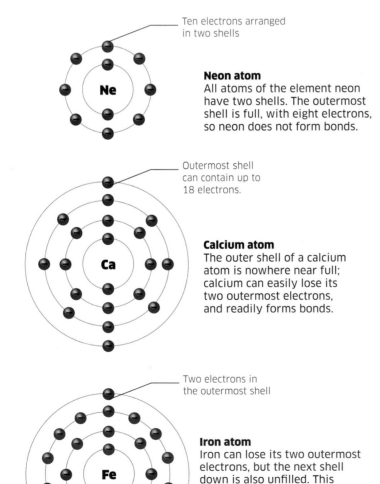

Ten electrons arranged in two shells

Neon atom
All atoms of the element neon have two shells. The outermost shell is full, with eight electrons, so neon does not form bonds.

Outermost shell can contain up to 18 electrons.

Calcium atom
The outer shell of a calcium atom is nowhere near full; calcium can easily lose its two outermost electrons, and readily forms bonds.

Two electrons in the outermost shell

Iron atom
Iron can lose its two outermost electrons, but the next shell down is also unfilled. This means that iron (and most other transition metals) can form all three types of bond – ionic, covalent, and metallic.

Ionic bonding

Many solids are made of ions: atoms, or groups of atoms, that carry an electric charge, because they have either more or fewer negative electrons than positive protons. Ions form when atoms (or groups of atoms) lose or gain electrons, in order to attain full outer electron shells. Electrical attraction between positive ions (+) and negative ions (-) causes the ions to stick together, forming a crystal.

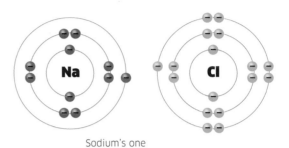

Two atoms
Neither sodium (Na) nor chlorine (Cl) atoms have filled outer shells. Sodium will easily give up its outermost electron.

Sodium's one outermost electron transfers

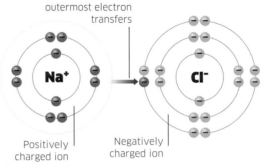

Electron transfer
Chlorine readily accepts the electron, so now both atoms have filled outer shells. They have become electrically charged and are now ions.

Positively charged ion

Negatively charged ion

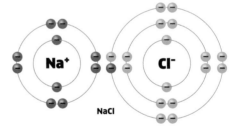

NaCl

Electrical attraction
The positive sodium ion and the negative chlorine ion are attracted to each other. They have become a compound called sodium chloride (NaCl).

Sodium ions and chlorine ions are held together by electrical attraction.

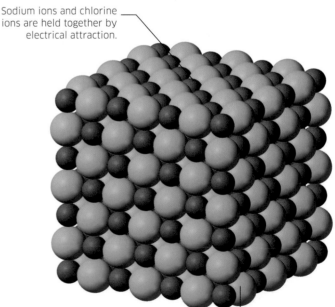

Ionic crystal
Ions of opposite electric charge are attracted to each other, and they form a regular pattern called a crystal. Many solids are ionic crystals, such as salt.

Salt crystal
The ions arrange in a regular pattern, forming a crystal of the compound sodium chloride (NaCl), or common salt.

The **atoms in your DNA** are held
together by **covalent bonds**.

4 The **number of bonds** each **carbon atom can form**, making it one
of the best atoms at **making up many different compounds**.

17

Covalent bonding

Another way atoms can attain full outer electron shells is by sharing electrons, in a covalent bond. A molecule is a group of atoms held together by covalent bonds (see pp.14–15). Some elements exist as molecules formed by pairs of atoms, for example chlorine, oxygen, and nitrogen. Covalent bonds can be single, double, or triple bonds.

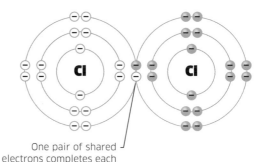

Single bond
Some pairs of atoms share only one electron each, forming a single bond.

One pair of shared electrons completes each chlorine atom's outer shell.

Each hydrogen atom needs one more electron to have a single full shell of two electrons.

The outermost shell of a nitrogen atom has five electrons so it is three electrons short of being full.

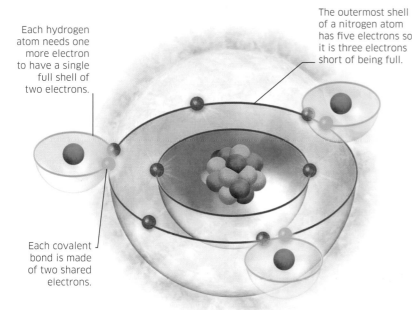

Each covalent bond is made of two shared electrons.

AMMONIA MOLECULE (NH₃)

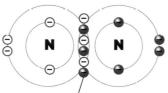

Double bond
Sometimes, pairs of atoms share two electrons each, forming a double bond.

Oxygen atoms share two electrons each to fill their outer shell.

Ammonia molecule
A molecule of the compound ammonia (NH_3) is made of atoms of nitrogen (N) and hydrogen (H). The shell closest to the nucleus of an atom can hold only two electrons. Hydrogen and helium are the only elements with just one shell.

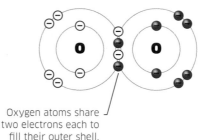

Before bonding, a nitrogen atom's outer shell is three electrons short, so it forms a triple bond.

Triple bond
In some pairs of atoms, three electrons are shared, forming a triple bond.

Metallic bonding

In a metal, the atoms are held in place within a "sea" of electrons. The atoms form a regular pattern – a crystal. Although the electrons hold the atoms in place, they are free of their atoms, and can move freely throughout the crystalline metal. This is why metals are good conductors of electricity and heat.

The metal changes shape as the hammer hits it.

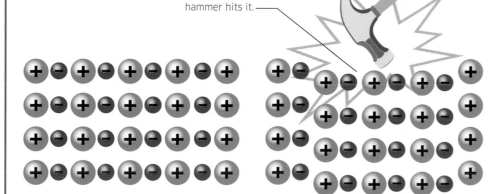

Conducting heat and electricity
An electric current is a flow of electric charge. In a metal, negatively charged electrons can move freely, so electric current can flow through them. The mobile electrons are also good at transferring heat within a metal.

Malleable metals
Metal atoms are held in place by metallic bonding, but are able to move a little within the "sea" of electrons. This is why metals are malleable (change shape when beaten with a hammer) and ductile (can be drawn into a wire).

Getting into shape
With some heat and a hammer, metals can be shaped into anything from delicate jewellery to sturdier objects, such as this horseshoe. Horseshoes used to be made of iron, but these days metal alloys such as steel (see p.63) are more common.

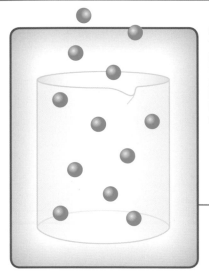

Gas state
The particles of a gas, such as oxygen, or the water vapour in the polar bear's breath, are not tightly held together by bonds. Without these forces keeping them together, they move freely in any direction.

Air
Air is a mixture of gases: mostly nitrogen (78 per cent), oxygen (21 per cent), and small proportions of argon and carbon dioxide.

Solids, liquids, and gases

There are four different states of matter: solid, liquid, gas, and plasma. Everything in the Universe is in one of those states. States can change depending on temperature and pressure.

All pure substances can exist in all of the three states common on Earth – solid, liquid, and gas. What state a substance is in is determined by how tightly its particles (atoms or molecules) are bound together. When energy (heat) is added, the tightly packed particles in a solid increase their vibration. With enough heat, they start moving around and the solid becomes a liquid. At boiling point, molecules start moving all over the place and the liquid becomes gas. Plasma is a type of gas so hot that its atoms split apart.

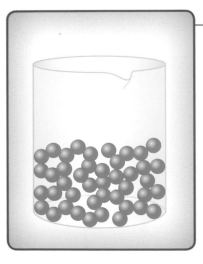

Liquid state
The particles of a liquid, such as water, are less tightly packed than in a solid and not neatly arranged, and they have weaker bonds. That is why liquids flow and spread, taking the shape of any container.

Saltwater
Salty seawater has a lower freezing point than freshwater, which freezes at 0°C (32°F). Because salt disrupts the bonds between water molecules, seawater stays liquid until about -2°C (28°F).

Water is one of the few substances that **expand** when **freezing**.

67 per cent of freshwater on Earth is in its solid state in the form of **ice caps and glaciers**.

19

Changing states of matter

Adding or removing energy (as heat) causes a state change. Solids melt into liquids, and liquids vaporize into gas. Some solids can turn straight to gas; some gases into solids.

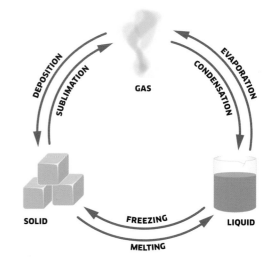

DEPOSITION
SUBLIMATION
GAS
EVAPORATION
CONDENSATION
SOLID
FREEZING
LIQUID
MELTING

Sublimation
Solid carbon dioxide is known as dry ice. With lowered pressure and increased heat it becomes CO_2 gas – this is called sublimation. When a gas goes straight to solid, the term is deposition.

Melting and freezing

All pure substances have a specific melting and freezing point. How high or low depends on how their molecules are arranged.

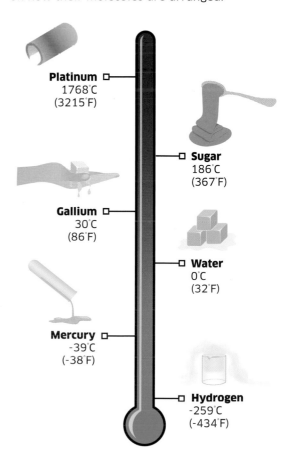

Platinum
1768°C
(3215°F)

Sugar
186°C
(367°F)

Gallium
30°C
(86°F)

Water
0°C
(32°F)

Mercury
-39°C
(-38°F)

Hydrogen
-259°C
(-434°F)

Plasma
Plasma, which makes up the Sun and stars, is the most common matter in the Universe. Intense heat makes its atoms separate into positively charged nuclei and negatively charged electrons that whizz about at very high speed.

Aurora borealis
Collisions between plasma from space and gases in the atmosphere energize atmospheric atoms, which release light when they return to normal energy levels.

States of matter

Water exists in three states. Here we see it as solid ice, liquid seawater, and gaseous water vapour, exhaled by the polar bear. Water vapour is invisible until it cools and condenses to form steam, a mist of liquid droplets – the same happens when a pan of water boils. In the Arctic Circle, the spectacular northern lights (aurora borealis) reveal the presence of plasma, the fourth state of matter.

Solid state
In a solid, such as ice, particles are held together by bonds and sit tightly packed. The particles vibrate slightly but they don't move about, so solids keep their shape.

Mixtures

When two or more substances are mixed together, but do not bond chemically to make a compound, they form a mixture. In a mixture, substances can be separated by physical means.

Mixtures are all around us, both natural and man-made. Air is a mixture of gases. Soil is a mixture of minerals, biological material, and water. The pages of this book are a mixture of wood pulp and additives, and the ink on the pages is a mixture of pigments. There are different types of mixtures. Salt dissolved in water is a solution. Grainy sand mixed with water forms a suspension. A colloid is a mix of tiny particles evenly dispersed, but not dissolved, in another substance; mist is a colloid of minute droplets of water in air. Evenly distributed mixtures are homogeneous, uneven mixtures are heterogeneous (see also pp. 10–11).

Spray
Sea spray is a heterogeneous mixture of air and seawater.

Salty solution
The saltwater in the sea is a solution: a homogeneous mixture of water and dissolved salts. When seawater evaporates, salt crystals are formed.

Sand
Sand is a heterogeneous mixture: a close look reveals tiny pieces of eroded rock, crushed shells, glass, and even bits of plastic.

Organic matter
Fish and other sea creatures release organic matter, such as poo and old scales, into the sea.

Mixtures in nature

Most substances in nature are mixtures, including seawater, rocks, soil, and air. Understanding how to separate these mixtures provides us with an important supply of natural resources, for example by removing salt from seawater and separating gases, such as argon, from air.

Seaweed
Dead and decaying algae also contribute organic matter to the seawater mix.

More than **5 million tonnes** of gold are dispersed as tiny particles in the world's oceans.

Seawater is an important source of the useful element **magnesium**, an alkaline earth metal.

21

Sea foam
Sea foam forms at the water's edge when wind and waves whip up air and water to frothy bubbles, which mix with biological material excreted from algae and other sealife.

Rock
Lots of different minerals can make up the solid mixture that forms rocks. Most of the minerals that are present in seawater come from eroded rock.

Seawater
The oceans are full of materials, dissolved as well as dispersed (scattered) in water: salts, gases, metals, organic compounds, and microscopic organisms. This type of uneven mixture is called a suspension.

Separating mixtures
There are many ways to separate mixtures, whether it is to extract a substance, or analyse a mixture's contents. Different techniques work for different substances, depending on their physical properties.

Filtration
Filtration separates insoluble solids from liquids, which pass through the filter.

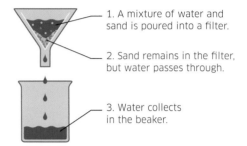

1. A mixture of water and sand is poured into a filter.

2. Sand remains in the filter, but water passes through.

3. Water collects in the beaker.

Chromatography
How fast substances in a liquid mixture such as ink separate, depends on how well they dissolve – the better they do, the further up the soaked paper they travel with the solvent.

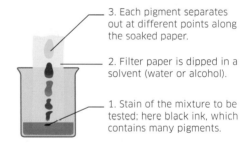

3. Each pigment separates out at different points along the soaked paper.

2. Filter paper is dipped in a solvent (water or alcohol).

1. Stain of the mixture to be tested; here black ink, which contains many pigments.

Distillation
This method separates liquids according to their boiling point. The mixture is heated, and the substance that boils first evaporates and can be collected as it condenses.

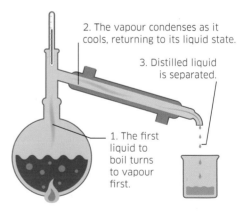

2. The vapour condenses as it cools, returning to its liquid state.

3. Distilled liquid is separated.

1. The first liquid to boil turns to vapour first.

Magnetism
Passing a magnet over a mixture of magnetic and non-magnetic particles removes the magnetic ones.

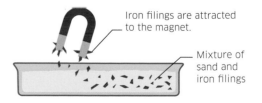

Iron filings are attracted to the magnet.

Mixture of sand and iron filings

Rocks and minerals

The chemistry of Earth is dominated by the huge variety of rocks and minerals that shape the landscape around us.

There are thousands of different kinds of rocks and minerals. What they are like depends on the chemical elements they contain, and the way these elements are grouped together. A rock is a mixture of different minerals, arranged as billions of tiny grains. Each mineral is usually a compound of two or more elements chemically bonded together. Many of these form beautiful crystals. Sometimes, a mineral is an element in its raw form – such as copper or gold.

Most of the ocean floor is made of
igneous basalt rock,
much younger than most rocks on land.

Sedimentary rock
Fragments of rock broken away by weathering and erosion join together to form sedimentary rocks, such as sandstone (below) and limestone. The fragments gather in layers at the bottom of lakes and oceans, and get compacted and cemented together under their own weight. Eventually, uplift pushes this rock up to the surface.

The rock cycle
Solid rocks look like they must stay the same for ever, but in fact they change over thousands or millions of years. Some melt under the influence of Earth's internal heat and pressure. Others get eroded by wind and rain. The three main forms of rock are linked in a cycle that changes one form into another. The cycle is driven slowly, but inevitably, by a set of dramatic movements deep within the Earth.

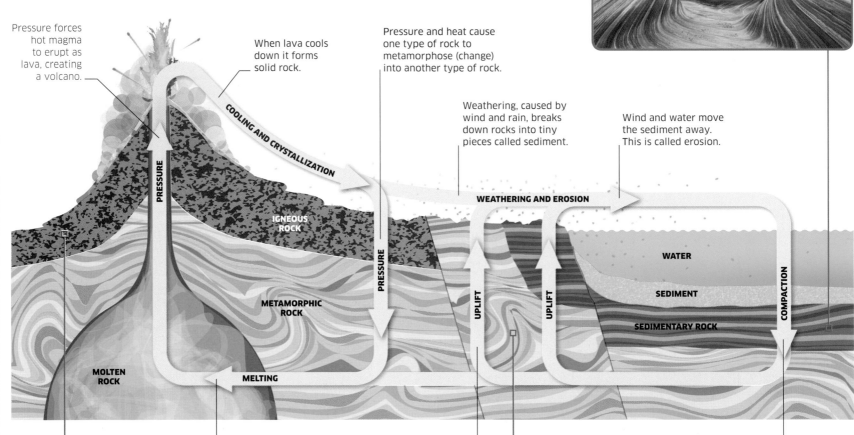

Pressure forces hot magma to erupt as lava, creating a volcano.

When lava cools down it forms solid rock.

Pressure and heat cause one type of rock to metamorphose (change) into another type of rock.

Weathering, caused by wind and rain, breaks down rocks into tiny pieces called sediment.

Wind and water move the sediment away. This is called erosion.

COOLING AND CRYSTALLIZATION

WEATHERING AND EROSION

PRESSURE

IGNEOUS ROCK

WATER

PRESSURE

METAMORPHIC ROCK

UPLIFT

UPLIFT

SEDIMENT

COMPACTION

SEDIMENTARY ROCK

MOLTEN ROCK

MELTING

Heat from deep underground melts solid rock to form liquid magma.

Rocks can move up to the surface as new rock forms underneath, a process known as uplift.

Layers of sediment settle, and then get compacted (squashed together) into sedimentary rock.

Igneous rock
The interior of the Earth is so hot it melts solid rock, forming a liquid called magma. When magma cools down it solidifies and crystallizes to form igneous rock, such as granite (formed underground) and basalt (seen left) from lava erupted from volcanoes.

Metamorphic rock
Rocks that get buried deep underground are squeezed and heated under pressure. But instead of melting the rock, this rearranges its crystals to form metamorphic rock. For example, buried limestone changes into marble, as in this cave.

Earth's upper mantle, just beneath the crust, consists mainly of **very hot peridotite**, a green igneous rock.

Dark green **imperial jade** is one of the **rarest and most precious** minerals in the world.

23

Elements of Earth's crust

Planet Earth is mostly made up of the elements iron, oxygen, silicon, and magnesium, with most of the iron concentrated in Earth's core. But Earth's outer layer, the crust, is made from minerals of many different elements, such as silicates (containing silicon and oxygen). This diagram shows which elements are most common in the crust.

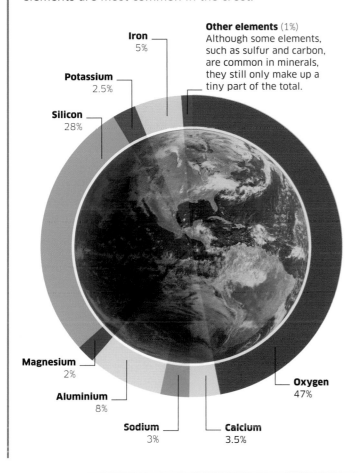

Other elements (1%)
Although some elements, such as sulfur and carbon, are common in minerals, they still only make up a tiny part of the total.

Iron 5%

Potassium 2.5%

Silicon 28%

Magnesium 2%

Aluminium 8%

Sodium 3%

Calcium 3.5%

Oxygen 47%

Native elements

In Earth's crust, most elements exist combined with others in mineral compounds. But some, called native elements, appear in pure form. About 20 elements can be found in pure form, including metals, such as copper and gold, and non-metals, such as sulfur and carbon.

Sulfur
Powder and crystals of pure sulfur from volcanic gases accumulate around volcanic vents. In the rock cycle, it gets mixed into rocks. It also forms part of many mineral compounds.

Mineral compounds

There are more than 4,000 different kinds of minerals. Scientists classify them according to which elements they contain, and sort them into a few main groups. The group name tells which is the main element in all minerals in that group. All sulfide minerals, for example, contain sulfur. Many minerals exist in ores – rocks from which metals can be extracted – or as pretty gem crystals (see p.24).

Hematite
This oxide contains lots of iron, making it an important iron ore.

Rose quartz
This is a pink form of quartz, one of the silicates made up of only silicon and oxygen.

Oxides
Different metals combine with oxygen to form these hard minerals. They are in many ores, making these valuable sources of metal. Many make fine gems.

Silicates
All silicates, the most common group, contain silicon and oxygen. Some include other elements, too. The rock granite is made of three silicates, including quartz.

Baryte
The element barium combined with sulphur and oxygen makes baryte, which comes in many different forms.

Chalcopyrite
Both copper and iron can be sourced from ores containing this sulfide.

Sulfates
A sulfur and oxygen compound combines with other elements to form sulfates. Most common are gypsum, which forms cave crystals (see pp.26–27), and baryte.

Sulfides
Metals combined with sulfur, but no oxygen, form sulfides. Sulfides make up many metal ores. Many are colourful, but are usually too soft to use as gemstones.

Malachite
Copper combines with carbon and oxygen to give this useful and decorative mineral its green colour.

Fluorite
Calcium and fluorine make up this mineral, which comes in many different colours.

Carbonates
Compounds of carbon and oxygen combine with other elements to form carbonates. Many are quite soft. Some exist in rocks such as chalk and limestone.

Halides
These minerals contain one or more metals combined with a halogen element (fluorine, chlorine, bromine, or iodine; see p.40). Rock salt is an edible halide.

Crystals

A crystal is a solid material, made of atoms set in a repeating 3D-pattern. Crystals form from minerals when molten magma cools to become solid rock. Crystals of some substances, such as salt, sugar, and ice, are formed through evaporation or freezing.

The shapes and colours of mineral crystals depend on the elements from which they are made, and the conditions (the temperature and pressure) under which they formed. The speed at which the magma cools decides the size of the crystals. Crystals can change under extreme pressure in the rock cycle (see p.22), when one rock type changes into another.

Crystal structures

Crystals have highly ordered structures. This is because the atoms or molecules in a crystal are arranged in a 3D-pattern that repeats itself exactly over and over again. Most metals have a crystalline structure, too.

Quartz tetrahedron
The molecule that makes up quartz is in the shape of a tetrahedron, made of four oxygen atoms and one silicon atom.

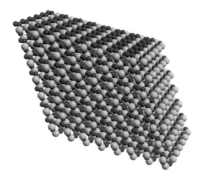

Quartz crystal
A quartz crystal consists of a lattice of tetrahedrons, repeated in all directions.

One mineral, two gem crystals

Crystals of the mineral corundum come in many colours, thanks to different impurities in the crystal structure. Often cut and polished to be used as gems, the best known are sapphire (usually blue) and ruby (red).

BLUE CORUNDUM: SAPPHIRE

RED CORUNDUM: RUBY

CUT RUBY CRYSTAL SET IN A RING

Quartz crystals

The crystal quartz is one of the most common minerals in Earth's crust. It comes in many different forms and colours – but they all share the same formula: silicon dioxide, or SiO_2. Some of the best known include rock crystal (transparent), rose quartz (pink), tiger-eye (yellow-brown), citrine (yellow), and amethyst (purple). Their beauty makes them popular for jewellery, whether in natural form, tumbled, or cut and polished.

Amethyst geode
A geode is formed when gas bubbles are trapped in cooling lava. The crystals lining the walls of the geode grow when hot substances containing silicon and oxygen, as well as traces of iron, seep into the cavities left by the bubbles.

The purple colour of amethyst comes from iron impurities in the crystal structure.

The outer shell of the geode is normally a volcanic, igneous rock such as basalt.

Crystal systems

The shape of a crystal is determined by how its atoms are arranged. This decides the number of flat sides, sharp edges, and corners of a crystal. Crystals are sorted into six main groups, known as systems, according to which 3D-pattern they fit.

Cubic
Gold, silver, diamond, the mineral pyrite (above), and sea salt all form cubic crystals.

Tetragonal
Zircon, a silicate mineral, is a typical tetragonal crystal, looking like a square prism.

Hexagonal and trigonal
Apatite is a hexagonal crystal, with six long sides. Trigonal crystals have three sides.

Monoclinic
Orthoclase (above) and gypsum crystals are monoclinic, one of the most common systems.

The **largest quartz crystal cluster** in the world is 3 m (9.8 ft) tall and weighs more than **14,000 kg (30,000 lb)**.

25

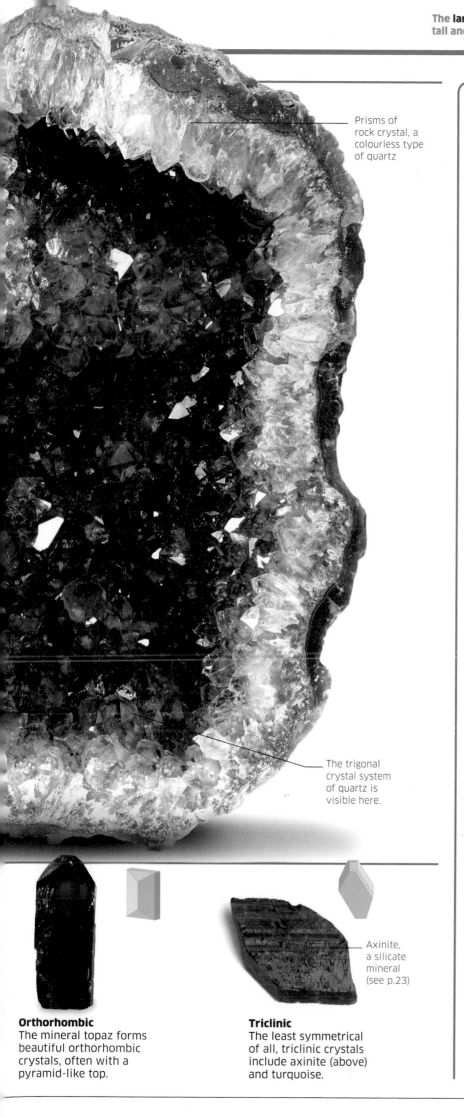

Prisms of rock crystal, a colourless type of quartz

The trigonal crystal system of quartz is visible here.

Orthorhombic
The mineral topaz forms beautiful orthorhombic crystals, often with a pyramid-like top.

Triclinic
The least symmetrical of all, triclinic crystals include axinite (above) and turquoise.

Axinite, a silicate mineral (see p.23)

Ice crystals
In an ice crystal, water molecules are aligned hexagonally. These crystals form when water vapour in the air freezes straight to a solid. If liquid water freezes slowly, it will form simple hexagonal crystals, but without the delicate branches and shapes of a snowflake crystal.

The unique pattern of a snowflake is based on a six-sided shape (hexagon).

Snowflake
A snowflake is a six-sided ice crystal. Each snowflake grows into a different variation on this shape, depending on how it drifts down from the sky. No two snowflakes are the same.

Sugar and salt crystals
Crystals of sea salt and crystals of sugar are more different than they look. Salt crystals are highly ordered six-sided cubes, while sugar crystals are less well ordered hexagonal prisms.

Sea salt belongs to the cubic crystal system, but when the crystals form quickly they take a pyramid shape.

Sea salt crystals
Crystals of sea salt (sodium chloride) are held together by ionic bonds (see p.16). When salt water evaporates, the dissolved minerals left behind form salt crystals.

Liquid crystals
In nature, cell membranes and the solution produced by silkworms to spin their cocoons are liquid crystals. The molecules in liquid crystals are highly ordered, but they flow like a liquid.

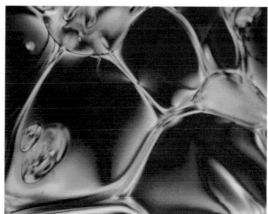

Liquid crystals at work
Man-made liquid crystals, such as the ones seen here, are used in liquid crystal displays (LCDs) in TV screens, digital watches, and mobile phones. They do not produce light, but create clear images by altering the way light passes through them.

Crystal cave

In extremely hot and humid conditions, these scientists are investigating the largest crystals ever found, in the Giant Crystal Cave, Naica, Mexico.

The crystals are made of selenite, a form of the mineral gypsum (calcium sulfate), which is the main ingredient of plaster and blackboard chalk. The crystals form very slowly, from calcium, sulfur, and oxygen dissolved in hot water. This water was heated by magma in a geological fault beneath the cave. The largest crystals weigh 50 tonnes and are 12 m (39 ft) long.

THE ELEMENTS

Shiny gold, tough iron, smelly chlorine, and invisible oxygen – what have they got in common? They are all elements: substances made of only one type of atom that cannot be broken down into a simpler substance. But they can combine with other elements to form new substances, known as compounds. Everything around us is made up of elements, either in pure form or combined. Water, for example, is made of the elements hydrogen and oxygen. Out of 118 known elements, around 90 exist naturally. The rest have been created in laboratory experiments.

Atomic number
This is the number of protons in the atom's nucleus. The element iron has an atomic number of 26, which means it has 26 protons (and 26 electrons).

Atomic mass number
An atom's mass is how many protons and neutrons it has. This number shows the relative atomic mass (the average mass of all an element's atoms, see p.13).

26 55.845
Fe
IRON

Name
In English, some element names look very different to their symbol. We say "iron" rather than "ferrum", its original Latin name.

Chemical symbol
An element has the same symbol all over the world, while the name can be different in different languages.

The periodic table
In 1869, the Russian scientist Dmitri Mendeleev came up with a system for how to sort and classify all the elements. In his chart, the atomic number increases left to right, starting at the top left with hydrogen, with an atomic number of 1. Arranging elements in rows and columns reveals patterns. For example, elements from the same column, or group, react in similar ways and form a part of similar compounds.

Elemental information
An element's place in the table is decided by its atomic number. Each element has a "tile" showing its atomic number, its chemical symbol, and its atomic weight (weights in brackets are estimates for unstable elements). The symbol is an abbreviation of the element's original name. This name was often invented by the person who discovered the element.

1

1											
1 1.0079 **H** HYDROGEN	**2**										
3 6.941 **Li** LITHIUM	**4** 9.0122 **Be** BERYLLIUM										
11 22.990 **Na** SODIUM	**12** 24.305 **Mg** MAGNESIUM	**3**	**4**	**5**	**6**	**7**	**8**	**9**	**10**	**11**	**12**
19 39.098 **K** POTASSIUM	**20** 40.078 **Ca** CALCIUM	**21** 44.956 **Sc** SCANDIUM	**22** 47.867 **Ti** TITANIUM	**23** 50.942 **V** VANADIUM	**24** 51.996 **Cr** CHROMIUM	**25** 54.938 **Mn** MANGANESE	**26** 55.845 **Fe** IRON	**27** 58.933 **Co** COBALT	**28** 58.693 **Ni** NICKEL	**29** 63.546 **Cu** COPPER	**30** 65.39 **Zn** ZINC
37 85.468 **Rb** RUBIDIUM	**38** 87.62 **Sr** STRONTIUM	**39** 88.906 **Y** YTTRIUM	**40** 91.224 **Zr** ZIRCONIUM	**41** 92.906 **Nb** NIOBIUM	**42** 95.94 **Mo** MOLYBDENUM	**43** (96) **Tc** TECHNETIUM	**44** 101.07 **Ru** RUTHENIUM	**45** 102.91 **Rh** RHODIUM	**46** 106.42 **Pd** PALLADIUM	**47** 107.87 **Ag** SILVER	**48** 112.41 **Cd** CADMIUM
55 132.91 **Cs** CAESIUM	**56** 137.33 **Ba** BARIUM	**57-71** **La-Lu** LANTHANIDES	**72** 178.49 **Hf** HAFNIUM	**73** 180.95 **Ta** TANTALUM	**74** 183.84 **W** TUNGSTEN	**75** 186.21 **Re** RHENIUM	**76** 190.23 **Os** OSMIUM	**77** 192.22 **Ir** IRIDIUM	**78** 195.08 **Pt** PLATINUM	**79** 196.97 **Au** GOLD	**80** 200.59 **Hg** MERCURY
87 (223) **Fr** FRANCIUM	**88** (226) **Ra** RADIUM	**89-103** **Ac-Lr** ACTINIDES	**104** (261) **Rf** RUTHERFORDIUM	**105** (262) **Db** DUBNIUM	**106** (266) **Sg** SEABORGIUM	**107** (264) **Bh** BOHRIUM	**108** (277) **Hs** HASSIUM	**109** (268) **Mt** MEITNERIUM	**110** (281) **Ds** DARMSTADTIUM	**111** (282) **Rg** ROENTGENIUM	**112** (285) **Cn** COPERNICIUM

Lanthanides and actinides
Periods 6 and 7 each contain 14 more elements than periods 4 and 5. This makes the table too wide to fit easily in books, so these elements are shown separately. All elements in the actinides group are radioactive.

57 138.91 **La** LANTHANUM	**58** 140.12 **Ce** CERIUM	**59** 140.91 **Pr** PRASEODYMIUM	**60** 144.24 **Nd** NEODYMIUM	**61** (145) **Pm** PROMETHIUM	**62** 150.36 **Sm** SAMARIUM	**63** 151.96 **Eu** EUROPIUM	**64** 157.25 **Gd** GADOLINIUM	**65** 158.93 **Tb** TERBIUM
89 (227) **Ac** ACTINIUM	**90** 232.04 **Th** THORIUM	**91** 231.04 **Pa** PROTACTINIUM	**92** 238.03 **U** URANIUM	**93** (237) **Np** NEPTUNIUM	**94** (244) **Pu** PLUTONIUM	**95** (243) **Am** AMERICIUM	**96** (247) **Cm** CURIUM	**97** (247) **Bk** BERKELIUM

Periodic table key

- ■ Alkali metals
- ■ Alkaline earth metals
- □ Transition metals
- ■ Lanthanide metals
- ■ Actinide metals
- ■ Other metals
- ■ Metalloids
- ■ Other non-metals
- ■ Halogens
- ■ Noble gases

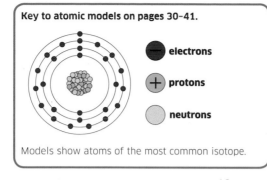

Key to atomic models on pages 30–41.

— electrons

+ protons

○ neutrons

Models show atoms of the most common isotope.

Other non-metals
These include three elements essential for life on Earth – carbon, nitrogen, and oxygen.

18

2	4.0026
He	
HELIUM	

13	14	15	16	17	

5 10.811 **B** BORON	6 12.011 **C** CARBON	7 14.007 **N** NITROGEN	8 15.999 **O** OXYGEN	9 18.998 **F** FLUORINE	10 20.180 **Ne** NEON
13 26.982 **Al** ALUMINIUM	14 28.086 **Si** SILICON	15 30.974 **P** PHOSPHORUS	16 32.065 **S** SULFUR	17 35.453 **Cl** CHLORINE	18 39.948 **Ar** ARGON
31 69.723 **Ga** GALLIUM	32 72.64 **Ge** GERMANIUM	33 74.922 **As** ARSENIC	34 78.96 **Se** SELENIUM	35 79.904 **Br** BROMINE	36 83.80 **Kr** KRYPTON
49 114.82 **In** INDIUM	50 118.71 **Sn** TIN	51 121.76 **Sb** ANTIMONY	52 127.60 **Te** TELLURIUM	53 126.90 **I** IODINE	54 131.29 **Xe** XENON
81 204.38 **Tl** THALLIUM	82 207.2 **Pb** LEAD	83 208.96 **Bi** BISMUTH	84 (209) **Po** POLONIUM	85 (210) **At** ASTATINE	86 (222) **Rn** RADON
113 (284) **Nh** NIHONIUM	114 (289) **Fl** FLEROVIUM	115 (288) **Mc** MOSCOVIUM	116 (293) **Lv** LIVERMORIUM	117 (294) **Ts** TENNESSINE	118 (294) **Og** OGANESSON

| 66 162.50 **Dy** DYSPROSIUM | 67 164.93 **Ho** HOLMIUM | 68 167.26 **Er** ERBIUM | 69 168.93 **Tm** THULIUM | 70 173.04 **Yb** YTTERBIUM | 71 174.97 **Lu** LUTETIUM |
| 98 (251) **Cf** CALIFORNIUM | 99 (252) **Es** EINSTEINIUM | 100 (257) **Fm** FERMIUM | 101 (258) **Md** MENDELEVIUM | 102 (259) **No** NOBELIUM | 103 (262) **Lr** LAWRENCIUM |

UNDERSTANDING THE PERIODIC TABLE

Within the table are blocks of elements that behave in similar ways. On the left are the most reactive metals. Most everyday metals occur in the middle of the table in a set called the transition metals. Non-metals are mostly on the right of the table and include both solids and gases.

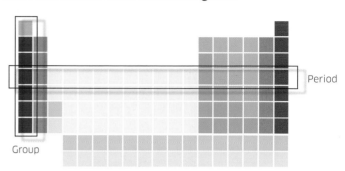

Period

Group

Building blocks

The periodic table is made up of rows called periods and columns called groups. As we move across each period, the elements change from solid metals (on the left) to gases (on the right).

Periods

All elements in a period have the same number of electron shells in their atoms. For example, all elements in the third period have three shells (but a different number of electrons).

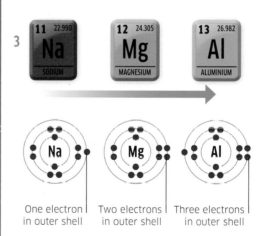

3

| 11 22.990 **Na** SODIUM | 12 24.305 **Mg** MAGNESIUM | 13 26.982 **Al** ALUMINIUM |

One electron in outer shell | Two electrons in outer shell | Three electrons in outer shell

Shrinking atoms
As you move along each row (period) of the table, the atoms of each element contain more protons and electrons. Each atom has the same number of electron shells, but for each step to the right, there are more positively charged protons pulling the shells inwards. This "shrinks" the atom, and makes it more tightly packed.

Groups

The elements in a group react in similar ways because they have the same number of electrons in their outer shell (see p.13). For example, while the elements in group 1 all have different numbers of electrons, and shells, they all have just one electron in their outer shell.

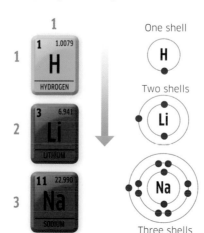

1

| 1 1.0079 **H** HYDROGEN | | 3 6.941 **Li** LITHIUM | | 11 22.990 **Na** SODIUM |

1
2
3

One shell — H
Two shells — Li
Three shells — Na

Growing atoms
Atoms get bigger and heavier as we move down each column (group). This is because the atoms of each element below have more protons and more electrons than the element above. As shells fill up with electrons (see p.13), a new shell is added each time we move another step down a group, down to the next period.

Transition metals

What we usually think of as "metals" mostly belong to the group of elements known as transition metals. Most are hard and shiny. They have many other properties in common, including high boiling points and being good at conducting heat and electricity.

The transition metals make up the biggest element block in the periodic table, spreading out from group 3 through to group 12, and across four periods (see pp.28–29). This wide spread indicates that, although they are similar in many ways, they vary in others, such as how easily they react and what kinds of compounds they form.

Some of these metals have been known for more than 5,000 years. Some were only discovered in the 20th century. This is a selection of some of the 38 transition metals.

SILVER
Argentum
Discovered: c.3000 BCE

Like gold and copper, silver is one of the elements known and used by the earliest civilizations. It is valuable and easy to mould and used to be made into coins. Today, coins are made of alloys (see pp.62–63). Silver is still one of the most popular metals for making jewellery and decorative objects.

Chunk of silver
Silver metal reacts with sulfur in air, which produces a black coating. That is why silver needs polishing to stay shiny.

Atomic structure
- ● 47
- ⊕ 47
- ○ 60

47 107.87
Ag
SILVER

OSMIUM
Osmium
Discovered: 1803

This rare, blue-shimmering metal is incredibly dense – a tennis ball-sized lump of osmium would have a mass of 3.5 kg (7.7 lb). If exposed to air, it reacts with oxygen to form a poisonous oxide compound, so for safe use it needs to be combined with other metals or elements. The powder used to detect fingerprints contains osmium.

Atomic structure
- ● 76
- ⊕ 76
- ○ 116

76 190.23
Os
OSMIUM

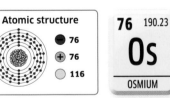

Hard but brittle
This sample of refined osmium looks solid enough, but the tiny cracks all over it show that it is fragile in its pure form.

GOLD
Aurum
Discovered: c.3000 BCE

Since ancient times, gold has been treasured because of its great beauty, and also because it doesn't get damaged by corrosion – it keeps its yellow sheen and does not rust. Easy to shape, it can be seen in jewellery, Egyptian masks, building decorations, and also in electronics. It doesn't easily react, or form compounds, with other elements.

Atomic structure
- ● 79
- ⊕ 79
- ○ 118

79 196.97
Au
GOLD

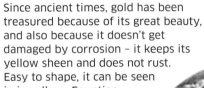
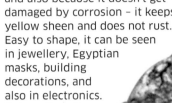
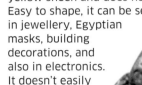

Gold nugget
In nature, pure gold can be found in nuggets such as this or, more commonly, as grains inside rocks.

Manganese is a transition metal which exists in tiny traces in **nuts and pineapples**.

The heavy transition metal **tungsten** has the **highest melting point** of any metal: **3,414°C (6,177.2°F)**.

31

COBALT
Cobaltum
Discovered: 1739

Cobalt is a metal somewhat similar to iron, its neighbour on the periodic table. It is often put in alloys, including those used to make permanent magnets. A cobalt compound has long been used to produce "cobalt blue", a deep vibrant blue for paints and dyes.

Atomic structure
- 27
+ 27
○ 32

27 58.933
Co
COBALT

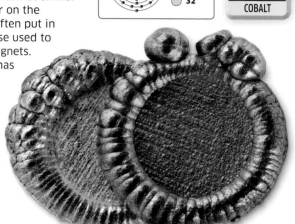

Cobalt colour
Extracted from its ore, pure cobalt metal is silvery grey in appearance.

CADMIUM
Cadmium
Discovered: 1817

Although it has some uses in industry and laser technology, this metal is now known to be highly toxic and dangerous to humans. If ingested, it can react like calcium, an essential and useful element, but will replace the calcium in our bones. This causes bones to become soft and easy to break.

Atomic structure
- 48
+ 48
○ 66

48 112.41
Cd
CADMIUM

Poisonous pellet
This sample of pure cadmium has been refined in a laboratory.

NICKEL
Niccolum
Discovered: 1751

This useful metal, which does not rust, is one of the ingredients in stainless steel (see p.63). It is also used to protect ships' propellers from rusting in water. Its best-known role is perhaps in the various alloys used to make coins, including the US 5-cent coin that is nicknamed "a nickel".

Atomic structure
- 28
+ 28
○ 30

28 58.693
Ni
NICKEL

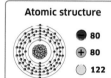

Pure nickel
These samples of pure nickel have been shaped into tiny balls.

MERCURY
Hydrargentum
Discovered: 1500 BCE

Famous for being the only metal that is liquid at room temperature, mercury has fascinated people for thousands of years. Only freezing to a solid at near -39°C (-38°F), it has long been used to measure temperature. But it is also poisonous, so thermometers now use other methods.

Atomic structure
- 80
+ 80
○ 122

80 200.59
Hg
MERCURY

Quick liquid
Mercury is also known as quicksilver, and it is easy to see why.

TITANIUM
Titanium
Discovered: 1791

Known for its strength, this metal was named after the Titans, the divine and tremendously forceful giants of Greek mythology. Titanium is hard but also lightweight, and resistant to corrosion. This super combination of properties makes it perfect for use in artificial joints and surgical pins, but also in watches and in alloys for the aerospace industry. It is, however, a very expensive material.

When titanium reacts with oxygen in the air it gets a duller grey coating. This actually works as a protection against corrosion.

Atomic structure
- 22
+ 22
○ 26

22 47.867
Ti
TITANIUM

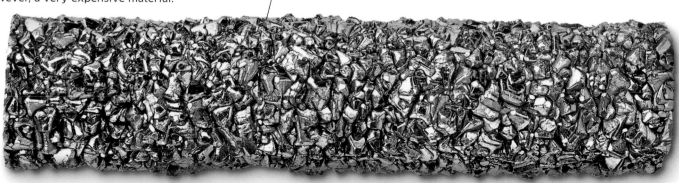

Laboratory sample
Although titanium is a common element in Earth's crust, it usually only exists in mineral compounds, not as a native element. Pure titanium has to be extracted and refined.

More metals

Most of the elements known to us are metals. In addition to the transition metals, there are five other metal groups in the periodic table, featuring a wide range of properties.

The alkali metals and alkaline earth metals are soft, shiny, and very reactive. The elements known as "other metals" are less reactive and have lower melting points. Underneath the transition metals are the lanthanides, which used to be called "rare earth metals", but turned out not to be rare at all, and the radioactive actinides. Whatever the group, these metals are all malleable, and good conductors of electricity and heat.

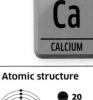

LITHIUM
Lithium
Discovered: 1817

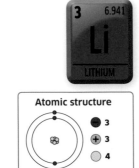

3 6.941
Li
LITHIUM

Lithium is the lightest of all metals. It has been used in alloys in the construction of spacecraft. In more familiar uses, we find lithium in batteries, but also in compounds used to make medicines.

Atomic structure
● 3
⊕ 3
○ 4

Pure lithium is a soft, silver-coloured metal.

SODIUM
Natrium
Discovered: 1807

11 22.990
Na
SODIUM

So soft it can easily be cut with a knife, and very reactive, sodium is more familiar to us when in compounds such as common salt (sodium chloride). It is essential for life, and plays a vital role in our bodies.

Atomic structure
● 11
⊕ 11
○ 12

Sodium is so reactive it needs to be stored away from air in sealed vials.

POTASSIUM
Kalium
Discovered: 1807

19 39.098
K
POTASSIUM

Along with sodium, the alkali metal potassium helps to control the nervous system in our bodies. We get it from foods such as bananas, avocados, and coconut water. It is added to fertilizers and is also part of a compound used in gunpowder.

Atomic structure
● 19
⊕ 19
○ 20

Highly reactive, potassium is often stored in oil to stop it reacting.

MAGNESIUM
Magnesium
Discovered: 1755

12 24.305
Mg
MAGNESIUM

Magnesium is an important metal because it is both strong and light in weight. The oceans are a main source of magnesium, but it is quite expensive to produce, so recycling it is crucial. As a powder, or thin strip, it is flammable and burns with a bright white light. It is often used in fireworks and flares.

Atomic structure
● 12
⊕ 12
○ 12

Magnesium is refined to produce a pure, shiny grey metal.

CALCIUM
Calcium
Discovered: 1808

20 40.078
Ca
CALCIUM

Our bodies are full of calcium, the fifth most common element on Earth. It makes teeth and bones strong, which is why it is important to eat calcium-rich foods, such as broccoli and oranges. It is also a vital part of compounds used to make cement and plaster.

Atomic structure
● 20
⊕ 20
○ 20

Pure metal samples such as this one are prepared using chemical processes. In nature, calcium is part of many minerals, but it doesn't exist on its own.

Aluminium is the most common metal in Earth's rocky crust.

Uranium, an actinide metal, was the first known radioactive element.

Atoms of the artificial element Moscovium break apart as soon as they have been made.

33

13 26.982
Al
ALUMINIUM

Atomic structure
- 13
+ 13
○ 14

ALUMINIUM
Aluminium
Discovered: 1825

Light and easy to shape, this metal is the main part of alloys used for anything from kitchen foil to aircraft parts. Much of it is recycled, as extracting it from mineral ores to produce pure metal is expensive and very energy-consuming.

TIN
Stannum
Discovered: c.3000 BCE

50 118.71
Sn
TIN

Tin was once smelted with copper to produce the alloy bronze – which led to the Bronze Age. Today it is used in alloys to plate other metal objects, such as pots and "tin cans".

Atomic structure
- 50
+ 50
○ 70

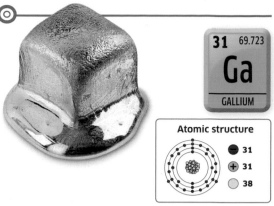

31 69.723
Ga
GALLIUM

Atomic structure
- 31
+ 31
○ 38

GALLIUM
Gallium
Discovered: 1875

Famous as an element with a melting point at just above room temperature, gallium metal melts in your hand. In commercial applications, gallium is a vital element in the production of semi-conductors for use in electronics.

THALLIUM
Thallium
Discovered: 1861

This soft, silvery metal is toxic in its pure state. It was commonly put to use as rat poison, but sometimes ended up killing humans too. Combined with other elements it can be useful, for example to improve the performance of lenses.

81 204.38
Tl
THALLIUM

Atomic structure
- 81
+ 81
○ 124

Toxic thallium in its pure form, safely kept in a vial

BISMUTH
Bismuthum
Discovered: 1753

Bismuth is a curious element. It is what is known as a heavy metal, similar to lead, but not very toxic. It is a tiny bit radioactive. It was not defined as an individual element until the 18th century, but has been known and used as a material since ancient times. For example in Egypt, at the time of the pharaohs, it added shimmer to make-up. It is still used in cosmetics today.

Atomic structure
- 83
+ 83
○ 126

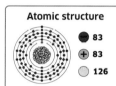

83 208.96
Bi
BISMUTH

Bismuth crystals
Brittle and grey in its pure metal form, bismuth can produce spectacular multicoloured crystals as an oxide compound.

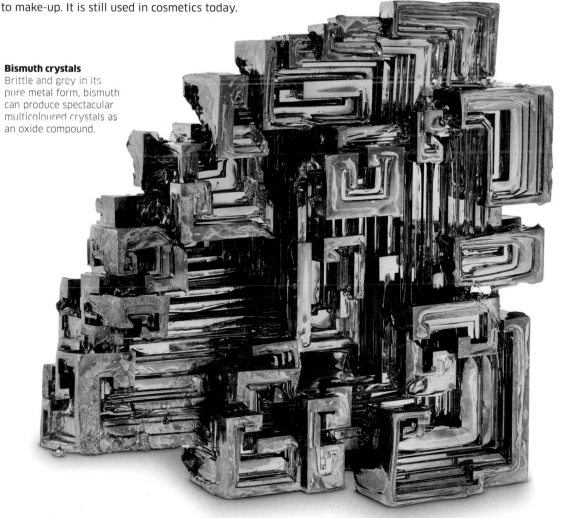

INDIUM
Indium
Discovered: 1863

A very soft metal in its pure state, indium is part of the alloy indium tin oxide, or ITO. This material is used in touch screens, LCD TV screens, and as a reflective coating for windows.

49 114.82
In
INDIUM

Atomic structure
- 49
+ 49
○ 66

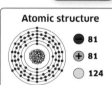

Metalloids

Also known as semi-metals, the metalloids are an odd collection of elements that show a wide range of chemical and physical properties. Sometimes they act as typical metals, sometimes like non-metals. One example of their behaviour as both is their use as semi-conductors in modern electronics.

In the periodic table, the metalloids form a jagged diagonal border between the metals on the left, and the non-metals to the right. Some scientists disagree regarding the exact classification of some elements in this part of the periodic table, precisely because of this in-between status. Some of the elements shown here are toxic, some are more useful than others, some are very common and some very rare. But they are all solid at room temperature.

Silicon sample
Pure silicon, such as this sample refined in a laboratory, shatters easily.

SILICON
Silicium
Discovered: 1823

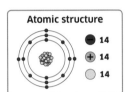

Atomic structure

● 14
⊕ 14
○ 14

14 28.086

Si

SILICON

Most of us are familiar with silicon, even if we don't know it. It is the second most abundant element on the Earth's crust, only after oxygen, and appears in many different silicate minerals. Mixed with other elements, silicon, a typical semi-conductor, is at the heart of the electronics industry – used in microchips and solar panels. Silicone baking moulds contain silicon, too.

Silicate minerals
Silicon is more or less everywhere, found in the silicate compounds that are better known to us as sand, quartz, talc, and feldspar, and in rocks made up of these minerals. Silicates also include minerals whose crystals make luxurious gems, such as amethyst, opal, lazurite, jade, and emerald. All these contain silica (silicon and oxygen), and sometimes other elements, too (see p.23).

Genesis rock
Collected on the Moon by Apollo 15 in 1971, this rock contains feldspar, a type of silicate mineral.

Moon mineral
It is not just on Earth that silicates abound. The surface of the Moon is made of 45 per cent silica.

Orthoclase
This feldspar is what gives pink granite its colour.

Feldspar minerals
A widespread group of silicate minerals, feldspars contain aluminium as well as silica, and often other elements too, including calcium, sodium, and potassium. They form common rocks, such as granite. The pretty crystal called moonstone is also a type of feldspar.

Silicate sands
Desert sand is chiefly composed of silica, a silicon and oxygen compound with the chemical name silicon dioxide. Sand started out as rock, that is gradually broken up and eroded into finer and finer grains. In the Sahara (left), this process started some 7 million years ago.

Tellurium is named after Tellus, the Latin name for planet Earth.

2 The number of Nobel Prizes won by Marie Curie, whose daughter Irene also won the Nobel Prize in chemistry.

35

BORON
Boron
Discovered: 1808

A hard element, boron gets even harder when combined with carbon as boron carbide. This is one of the toughest materials known, used in tank armour and bullet-proof vests. Boron compounds are used to make heat-resistant glass.

Atomic structure
● 5
⊕ 5
○ 6

5 10.811
B
BORON

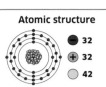

Dark and twisted
Pure boron is extracted from minerals in the deserts of Death Valley, USA.

GERMANIUM
Germanium
Discovered: 1886

In the history of the periodic table, germanium is an important element. In 1869, in his first table, Mendeleev predicted that there would be an element to fill a gap below silicon. It was discovered 17 years later, and did indeed fit there. Today germanium is used together with silicon in computer chips.

Atomic structure
● 32
⊕ 32
○ 42

32 72.64
Ge
GERMANIUM

Pure germanium
Refined germanium is shiny but brittle.

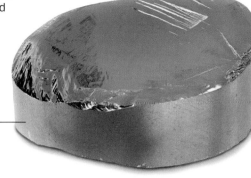

ARSENIC
Arsenicum
Discovered: 1250

Arsenic is an element with a deadly reputation. Throughout history, it has been used to poison people and animals, in fiction as well as in real life. Oddly, in the past it has been used as a medicine, too. It is sometimes used in alloys to strengthen lead, a soft, poisonous metal.

Atomic structure
● 33
⊕ 33
○ 42

33 74.922
As
ARSENIC

Dark matter
Pure arsenic can be refined from mineral compounds.

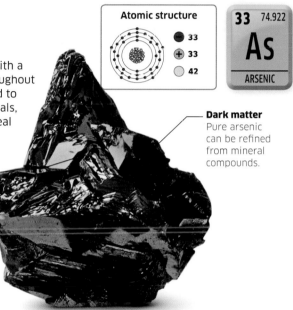

ANTIMONY
Stibium
Discovered: 1600 BCE

Antimony comes from stibnite, a naturally occurring mineral that also contains sulfur. Stibnite used to be ground up and made into eye makeup by ancient civilizations, as seen on Egyptian scrolls and death masks. Known as kohl, its Arabic name, it is still used in cosmetics in some parts of the world.

Atomic structure
● 51
⊕ 51
○ 70

51 121.76
Sb
ANTIMONY

Brittle crystals
This laboratory sample of refined antimony is hard but easily shattered.

TELLURIUM
Tellurium
Discovered: 1783

A rare element, in nature tellurium exists in compounds with other elements. It has a few specialist uses. It is used in alloys to make metal combinations easier to work with. It is mixed with lead to increase its hardness, and help to prevent it being damaged by acids. In rubber manufacture, it is added to make rubber objects more durable.

Atomic structure
● 52
⊕ 52
○ 78

52 127.60
Te
TELLURIUM

Refined tellurium
Silvery crystals of tellurium are often refined from by-products of copper mining.

POLONIUM
Polonium
Discovered: 1898

This highly radioactive and toxic element will forever be associated with the great scientist Marie Curie. Along with her husband Pierre, she discovered the element while researching radioactivity. She named it after her native Poland.

Atomic structure
● 84
⊕ 84
○ 125

84 (209)
Po
POLONIUM

Uraninite
Tiny amounts of polonium exist in this uranium ore.

Solid non-metals

Unlike metals, most non-metals do not conduct heat or electricity, and are known as insulators. They have other properties that are the opposite of those of metals, too, such as lower melting and boiling points.

On the right side of the periodic table are the elements that are described as non-metals. These include the halogens and the noble gases (see pp.40–41). There is also a set known as "other non-metals", which contains the elements carbon, sulfur, phosphorus, and selenium, all solids at room temperature. All of these exist in different forms, or allotropes. The "other non-metals" set of elements also includes a few gases (see pp.38–39).

Raw graphite
The surface of pure graphite looks metallic but is soft and slippery.

Raw diamond
Formed deep underground, raw diamonds are found in igneous (volcanic) rocks.

A clear diamond crystal like this can be cut into a precious gem.

PHOSPHORUS
Phosphorus
Discovered: 1669

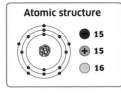

Atomic structure
- ● 15
- ⊕ 15
- ○ 16

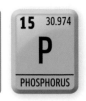
15 30.974
P
PHOSPHORUS

As a German alchemist boiled urine to produce the mythical philosopher's stone, he discovered a glowing, and very reactive, material instead. He named it phosphorus. It has a number of forms. The two most common are known as red phosphorus and white phosphorus.

Red phosphorus
More stable than white phosphorus, this form is used in safety matches, and fireworks.

White phosphorus
White phosphorus needs to be stored in water because it bursts into flames when in contact with air. It can cause terrible burns.

CARBON
Carbonium
Discovered: Prehistoric times

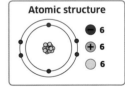

Atomic structure
- ● 6
- ⊕ 6
- ○ 6

6 12.011
C
CARBON

Carbon is at the centre of all life. This element forms the backbone of almost all the most important biological molecules. DNA, amino acids, proteins, fats, and sugars all contain multiple joined carbon atoms, bonded with other atoms, to form the molecules that make living organisms work. Carbon is in our bodies, in our food, in plants, and in most fuels we use for heating and transport. It appears as crystal-clear diamond as well as soft graphite.

Carbon allotropes

Allotropes are different forms of the same element. Carbon has three main allotropes: diamond, graphite, and buckminsterfullerene. It is the way the carbon atoms are arranged and bonded that determines which allotropes exist, and what their chemical and physical properties are.

Diamond
Diamond, an extremely hard allotrope of carbon, has its atoms arranged in a three-dimensional, rigid structure, with very strong bonds holding all of the atoms together.

Graphite
The "lead" in pencils is actually clay mixed with graphite, an allotrope in which the atoms bond in layers of hexagons. These can slide over each other, making it soft and greasy.

The **largest rough diamond** ever found, mined in South Africa, was just over **10 cm (4 in)** long.

The **Brazil nut** is the richest source of the form of **selenium** that the **human body needs**.

37

SULFUR
Sulfur

Discovered: 1777

This element has a distinctive yellow colour. Many compounds containing sulfur have a strong smell – for example, in rotten eggs and when onions are cut, it is sulfur that is at work. In ancient times it was known as brimstone, but it was only in 1777 that the French scientist Antoine Lavoisier discovered that it was in fact an element.

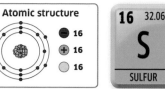

Atomic structure

● 16
⊕ 16
○ 16

16 32.065
S
SULFUR

Sulfur crystals
Crystals such as these can be found near volcanoes and hot springs (see p.23).

SELENIUM
Selenium

Discovered: 1817

Named after the Greek word *selene*, meaning "moon", selenium exists in three forms: red, grey, and black selenium. This is an element we need in just the right amount for our bodies to stay healthy, and it is a useful ingredient in anti-dandruff shampoo, but in some compounds it can be very toxic.

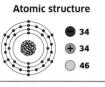

Atomic structure

● 34
⊕ 34
○ 46

34 78.96
Se
SELENIUM

Grey selenium
The most stable form of pure selenium is hard and shiny.

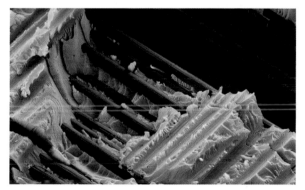

Carbon fibre
In modern materials technology, carbon fibres that are one-tenth of hair in thickness, but very tough, can be used to reinforce materials such as metals, or plastic (as seen above, enlarged many times).

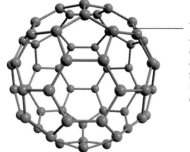

The carbon atoms are arranged in a rigid, stable structure that looks like a football.

Buckminsterfullerene
Nicknamed "buckyballs", buckminsterfullerene is any spherical molecule of carbon atoms, bonded in hexagons and pentagons. There are typically 60 atoms in a "ball". They exist in soot, but also in distant stars, and were only discovered in 1985.

Carbon fossil fuels
The substances we call hydrocarbon or fossil fuels include coal, natural gas, and oil. These fuels were formed over millions of years from decaying dead organisms. They are made up mainly of carbon and hydrogen, and when they burn they produce carbon dioxide gas (see p.50–51).

Coal
A long, slow process turned trees that grew on Earth some 300 million years ago into coal that we can mine today. As dead trees fell, they started to sink deep down in boggy soil. They slowly turned into peat, a form of dense soil, which can be burned when dried. Increasing heat and pressure compacted the peat further, turning it into lignite, a soft, brown rock. Even deeper down, the intense heat turned the lignite into solid coal.

Oil and natural gas
The crude oil that is used to make diesel and petrol is known as petroleum, meaning "oil from the rock". Millions of years ago, a layer of dead microorganisms covered the seabeds. It was slowly buried under mud and sand, gradually breaking down into hydrocarbons. Heat and pressure changed mud into rock and organic matter into liquid, or gas. This bubbled upwards until it reached a "lid" of solid rock, and an oil (or gas) field was formed.

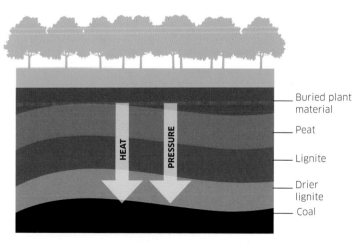

HEAT PRESSURE

Buried plant material

Peat

Lignite

Drier lignite

Coal

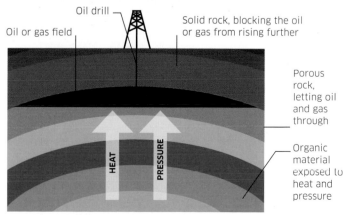

Oil drill

Oil or gas field

Solid rock, blocking the oil or gas from rising further

Porous rock, letting oil and gas through

Organic material exposed to heat and pressure

HEAT PRESSURE

Hydrogen, oxygen, and nitrogen

Among the non-metal elements, these three gases are vital to us in different ways. A mixture of nitrogen and oxygen makes up most of the air we breathe, while hydrogen is the most abundant element in the Universe.

Each of these gases has atoms that go in pairs: they exist as molecules of two atoms. That is why hydrogen is written as H_2, oxygen as O_2, and nitrogen as N_2. All three elements are found in compounds, such as DNA and proteins, that are vital for all forms of life on Earth.

HYDROGEN
Hydrogenium
Discovered: 1766

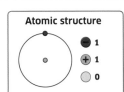

Atomic structure

● 1
⊕ 1
○ 0

1   1.0079
H
HYDROGEN

Hydrogen is the simplest of all the elements. Its lightest, and most common, isotope has atoms made of a single proton and a single electron, but no neutrons. Hydrogen gets its name from the Greek *hydro* and *genes* meaning "water forming"; when it reacts with oxygen it makes water, or H_2O.

Hydrogen in the Universe
Although rare in Earth's atmosphere, hydrogen makes up more than 88 per cent of all matter in the Universe. Our Sun is not much more than a ball of very hot hydrogen. The hydrogen fuses together to produce helium (see p.41), the second element in the periodic table. In the process, a vast amount of energy is produced.

Hydrogen as fuel
A very reactive element that will burn easily, hydrogen can be used as a fuel. When mixed with oxygen, it forms an explosive mixture. The rocket of a spacecraft uses liquid hydrogen, mixed together with liquid oxygen, as fuel. In fuel cells, used in electric cars, the chemical reaction between hydrogen and oxygen is converted to electricity. This combustion reaction produces only water, not water and carbon dioxide as in petrol-fuelled engines, making it an environmentally friendly fuel.

NITROGEN
Nitrogenium
Discovered: 1772

Atomic structure

● 7
⊕ 7
○ 7

In a nitrogen molecule (N_2), the two atoms are held together with a strong triple bond. The molecule is hard to break apart, which means nitrogen does not react readily with other substances. It is a very common element, making up 78 per cent of the air on Earth. It is extremely useful, too. We need it in our bodies and, as part of the nitrogen cycle (see p.186), it helps plants to grow. Where plants and crops need extra help, it is added to fertilizers.

7   14.007
N
NITROGEN

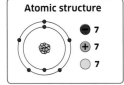

Explosive stuff
Molecules of nitrogen are not reactive, but many compounds containing nitrogen react very easily. These are found in many explosives, such as TNT, dynamite, and gunpowder, and in fireworks, too. On its own, compressed nitrogen gas is used to safely but powerfully blast out paintballs in paintball guns.

Liquid nitrogen
Nitrogen only condenses to liquid if it is cooled to -196°C (-321°F). This means that it is extremely cold in liquid form, instantly freezing anything it comes into contact with. This is useful for storing sensitive blood samples, cells, and tissue for medical use.

Jupiter is covered in **seas of liquid hydrogen**, formed as the **hydrogen in its atmosphere** condenses.

As a gas, oxygen is transparent, but in **its liquid form** it is pale blue.

39

OXYGEN
Oxygenium

Discovered: 1774

The element that we depend on to stay alive, oxygen was only recognized as an element in the late 18th century. Many chemists from different countries had for years been trying to work out precisely what made wood burn, and what air was made of, and several came to similar conclusions at roughly the same time. Oxygen is useful to us in many different forms and roles, some of which are described here.

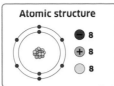

Atomic structure

- 8
- + 8
- 8

8   15.999
O
OXYGEN

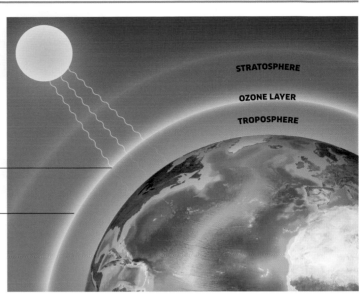

STRATOSPHERE

OZONE LAYER

TROPOSPHERE

Most of the harmful UV radiation from the Sun is absorbed by the ozone layer.

The ozone layer encircles Earth at a height of around 20 km (65,000 ft).

Fire
Three things are required for a fire to burn: there must be fuel, a source of heat such as a match, and oxygen gas. Without oxygen, no combustion (burning) can take place. Some fire extinguishers spray a layer of foam on the fire to prevent oxygen feeding it.

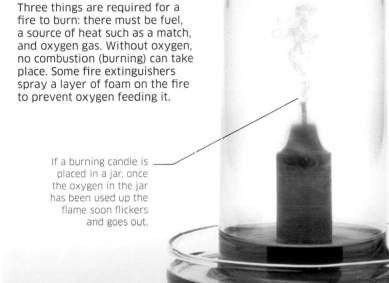

If a burning candle is placed in a jar, once the oxygen in the jar has been used up the flame soon flickers and goes out.

Air
Approximately 21 per cent, or one fifth, of the air in Earth's atmosphere is oxygen gas. In the lower atmosphere, the oxygen we breathe is the most common form of oxygen – molecules made up of two oxygen atoms (O_2). Higher above us, however, is the ozone layer that protects us from harmful ultraviolet rays from the Sun. Ozone (O_3) is another form, or allotrope, of oxygen, with three oxygen atoms in its molecules.

Life
All animals need oxygen to break down food and produce energy in a vital process called respiration.

Life on Earth
Our planet is the only one that has oxygen in its atmosphere. This is necessary for us to breathe. Oxygen is produced by photosynthesis, the process by which plants produce the food they need to live and grow. Water, which enabled life in the first place, millions of years ago, and is crucial to the survival of life in all forms, also contains oxygen. Even the ground is full of oxygen, in the form of different mineral compounds (see pp.22–23).

Land
Most of Earth's crust is made of rocks that contain oxygen compounds, such as these granite boulders.

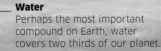

Water
Perhaps the most important compound on Earth, water covers two thirds of our planet.

CHEMICAL REACTIONS

A chemical reaction is what happens when one substance meets and reacts with another and a new substance is formed. The substances that react together are called reactants, and those formed are called products. In a chemical reaction, atoms are only rearranged, never created or destroyed.

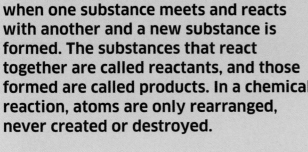

A change in colour of a substance often indicates that a reaction has happened.

DIFFERENT REACTIONS

There are many types of reaction. They vary depending on the reactants involved and the conditions in which they take place. Some reactions happen in an instant, and some take years. Exothermic reactions give off heat while endothermic reactions cool things down. The products in a reversible reaction can turn back into the reactants, but in an irreversible reaction they cannot. Redox reactions involve two simultaneous reactions: reduction and oxidation.

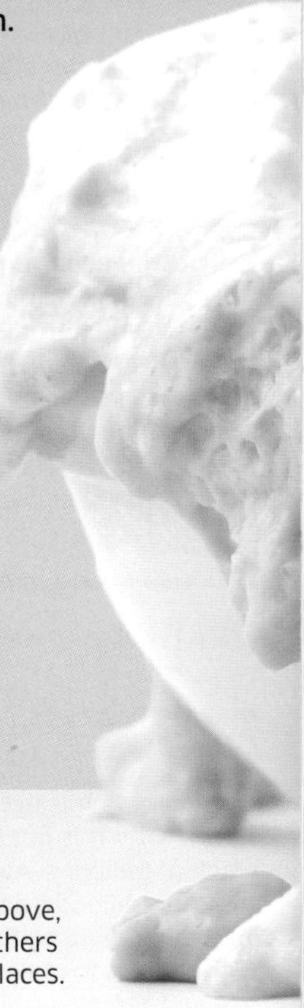

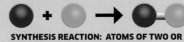

SYNTHESIS REACTION: ATOMS OF TWO OR MORE REACTANTS JOIN TOGETHER

DECOMPOSITION REACTION: ATOMS OF ONE REACTANT BREAKS APART INTO TWO PRODUCTS

DISPLACEMENT REACTION: ATOMS OF ONE TYPE SWAP PLACES WITH THOSE OF ANOTHER, FORMING NEW COMPOUNDS

Three kinds of reaction

Reactions can be classified in three main groups according to the fate of the reactants. As shown above, in some reactions the reactants join together, in others they break apart, and in some their atoms swap places.

REACTION BASICS

Chemical reactions are going on around us all the time. They help us digest food, they cause metal to rust, wood to burn, and food to rot. Chemical reactions can be fun to watch in a laboratory – they can send sparks flying, create puffs of smoke, or trigger dramatic colour changes. Some happen quietly, however, without us even noticing. The important fact behind all these reactions is that all the atoms involved remain unchanged. The atoms that were there at the beginning of the reactions are the same as the atoms at the end of the reaction. The only thing that has changed is how those atoms have been rearranged.

Reactants and products

The result of a chemical reaction is a chemical change, and the generation of a product or products that are different from the reactants. Often, the product looks nothing like the reactants. A solid might be formed by two liquids, a yellow liquid might turn blue, or a gas might be formed when a solid is mixed with a liquid. It doesn't always seem as if the atoms in the reactants are the same as those in the products, but they are.

Bonds are broken and reformed.

REACTANT 1   REACTANT 2   REACTION   PRODUCT

Chemical equations

The "law of conservation of mass" states that mass is neither created nor destroyed. This applies to the mass of the atoms involved in a reaction, and can be shown in a chemical equation. Reactants are written on the left, and products on the right. The number of atoms on the left of the arrow always equal those on the right. Everything is abbreviated: "2 H$_2$" means two molecules of hydrogen, with two atoms in each molecule.

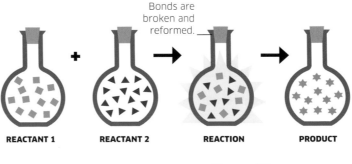

2 HYDROGEN MOLECULES (2 H$_2$)   1 OXYGEN MOLECULE (O$_2$)   2 WATER MOLECULES (2 H$_2$O)

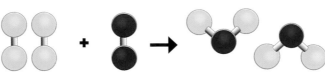

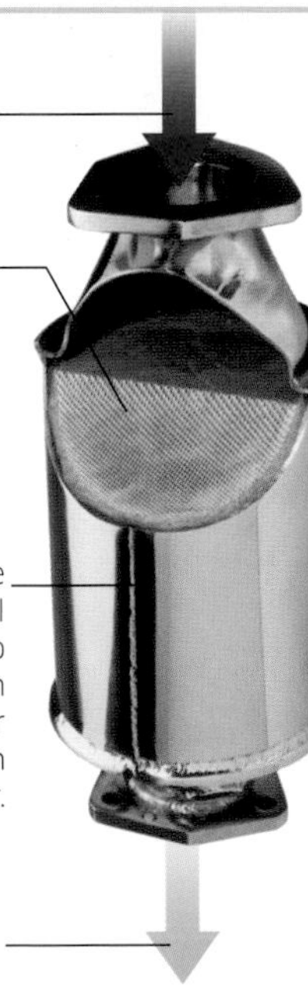

Dirty exhaust in

A car's catalytic converter contains a catalyst made of platinum and rhodium.

Carbon monoxide and unburnt fuel are converted to harmless carbon dioxide and water as they pass through the converter.

Cleaner exhaust out

Catalysts

Catalysts are substances that make chemical reactions go faster. Some reactions can't start without a catalyst. Catalysts help reactants interact, but they are not part of the reaction and remain unchanged. Different catalysts do different jobs. Cars use catalysts that help reduce harmful engine fumes by speeding up their conversion to cleaner exhausts.

Quick or slow?
Bread dough made with yeast rises slowly through fermentation. In this process, chemical compounds in the yeast react with sugar to produce bubbles of carbon dioxide gas, which make the dough rise. With baking soda, the reaction is between an acid and an alkali, which generates carbon dioxide in an instant.

Hot or cold?
It takes energy to break the bonds between atoms, while energy is released when new bonds form. Often, more energy is released than it takes to break the bonds. That energy is released as heat, such as when a candle burns. This is an exothermic reaction. If the energy released is less than the energy required to break the bonds, the reaction takes energy from its surroundings and both become colder. That reaction is endothermic.

Redox reactions
Redox reactions involve reduction (the removal of oxygen, or addition of electrons) and oxidation (the addition of oxygen or removal of electrons). When an apple turns brown in the air, a chemical inside the apple is oxidized, and oxygen from the air is reduced.

Reversible or irreversible?
Rusting is a redox reaction that, like an apple going brown, is irreversible. In a reversible reaction, certain products can turn back into their original reactants.

WHY DO REACTIONS HAPPEN?

Different chemical reactions happen for different reasons, including the type and concentration of reactants, temperature, and pressure. Chemical reactions involve the breaking and making of bonds between atoms. These bonds involve the electrons in the outer shell of each atom. It is how the electrons are arranged in atoms of different elements that decides which atoms can lose electrons and which ones gain them.

Why do atoms react?
Atoms that can easily lose electrons are likely to react with atoms that need to fill their outer shell. There are different types of bonds depending on how the atoms do this: covalent, ionic, and metallic (see pp.16–17). A water molecule (below) has covalent bonds.

By sharing electrons, hydrogen atoms get two electrons in their outer shell, and the oxygen outer shell is also full, with eight.

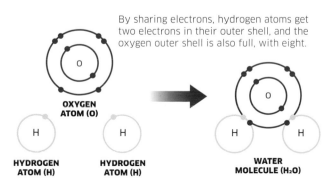

OXYGEN ATOM (O) · **HYDROGEN ATOM (H)** · **HYDROGEN ATOM (H)** · **WATER MOLECULE (H₂O)**

Increasing reactivity

POTASSIUM — Reacts with water
SODIUM
CALCIUM
MAGNESIUM
ALUMINIUM — Reacts with diluted acids
ZINC
IRON
COPPER
SILVER
GOLD — Hardly reacts at all

Metal reactivity series
A reactivity series sorts elements according to how readily they react with other elements. The most reactive is at the top; the least reactive at the bottom. It helps predicting how elements will behave in some chemical reactions.

Potassium
Potassium is the most reactive metal in the series. Adding a lump of potassium to water causes the potassium to react instantly: it whizzes around on the surface of the water and bursts into spectacular flames.

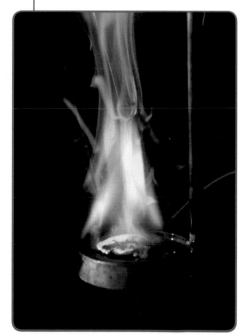

Compounds

When two or more elements join together by forming chemical bonds they make up a new, different substance. This substance is known as a compound.

Compounds are not just mixtures of elements. A mixture can be separated into the individual substances it contains, but it is not easy to turn a compound back into the elements that formed it. For example, water is a compound of hydrogen and oxygen. Only through a chemical reaction can it be changed back into these separate elements. A compound is made of atoms of two or more elements in a particular ratio. In water, for example, the ratio is two hydrogen atoms and one oxygen atom for every water molecule.

Fascinating formula

The chemical formula of a compound tells you which elements are present, and in what ratio. The compound sulfuric acid (H_2SO_4) is made of molecules that each contain two hydrogen atoms, one sulfur atom, and four oxygen atoms.

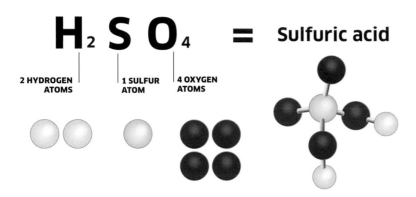

$$H_2 \, S \, O_4 \, = \, \textbf{Sulfuric acid}$$

2 HYDROGEN ATOMS   **1 SULFUR ATOM**   **4 OXYGEN ATOMS**

Great ways to bond

There are two types of bond that can hold the atoms in a compound together: covalent and ionic (see pp.16–17). Covalent bonds form between non-metal atoms. Ionic bonds form between metal and non-metal atoms.

Covalent compounds
Covalent compounds, such as sugar, form molecules in which the atoms form covalent bonds. They melt and boil at lower temperatures than ionic compounds. When they dissolve in water, they do not conduct electricity.

Salt lowers the freezing point of water, so it is used for melting ice and snow on roads.

Ionic compounds
Ionic compounds consist of ions. An ion is an electrically charged particle, formed when an atom has lost or gained electrons. Ions bond together, forming crystals with high melting points. Salt is an ionic compound.

Calcium carbonate is found in egg shells, but also in harder seashells.

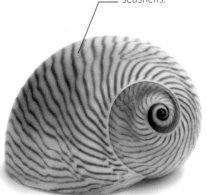

Best of both
Most compounds combine ionic and covalent bonding. In calcium carbonate, for example, calcium ions form ionic bonds with carbonate ions. Each carbonate ion contains carbon and oxygen atoms held together by covalent bonds.

Nothing like their elements

When atoms of different elements join to make new compounds, it is hard to tell what these elements are from looking at the compound. For example, no carbon is visible in carbon dioxide (CO_2), and no sodium in table salt, or sodium chloride (NaCl).

Salt, which contains the elements sodium and chlorine, looks nothing like either.

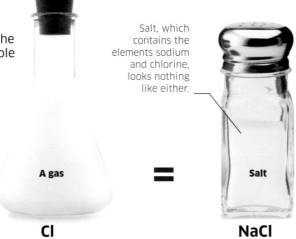

A metal   **+**   **A gas**   **=**   **Salt**

Na
Sodium

Cl
Chlorine

NaCl
Sodium chloride

Look what they have become
In chemical reactions, atoms from different elements regroup into new, different atom combinations. The resulting substances often look, and feel, completely different, too. For instance, sodium is a shiny metal, and chlorine is a pale green gas, but together they make sodium chloride (salt), a white crystal.

Pyrite, a form of iron sulfide

Iron sulfide
Iron sulfide, a compound of iron and sulfur, exists in several forms. Iron filings and yellow sulfur powder can be fused together to form a black solid called iron (II) sulfide (FeS). The mineral pyrite (FeS_2, above), known as "fool's gold", is another form of iron sulfide. Unlike iron, neither of these compounds is magnetic.

1862 The year the first plastic, Parkesine, was presented to the public. It was used to make buttons.

Cellulose, a natural polymer, is used to make cellophane, often used in sweet wrappers.

45

Polymers

Some molecules join up in a chain to form long polymers (meaning "many parts"). The smaller molecules that make up the polymer are called monomers. There are many important polymers in living things. Cellulose, which makes up wood, is the most abundant natural polymer on Earth. The DNA in our bodies, and starch in foods such as pasta, rice, and potatoes are also polymers. Polymers can be man-made, too. Synthetic polymers include a vast array of different plastics.

Plastic polymers and recycling
The first man-made polymers were attempts to reproduce the natural polymers silk, cellulose, and latex (see pp.58–59). Today, plastics play a massive role in the way we live, but they also pose a serious risk to the environment. In 1988, an identification code was developed to make plastics recycling easier. The code's symbols let the recyclers know what plastic an object is made of, which matters when it comes to process and recycle it.

What makes a polymer
A polymer is like a long string of beads, with each bead, or monomer, in the string made up of exactly the same combination of atoms. Shorter ones, with just two monomers, are called dimers, while those with three are known as trimers.

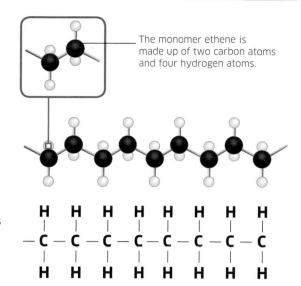

The monomer ethene is made up of two carbon atoms and four hydrogen atoms.

Polyethylene polymer
A string of ethene monomers is known as polyethylene (or polyethene/polythene). There are several thousand ethene monomers in a polyethylene polymer.

Type of plastic	Symbol	Properties	Use
Polyethylene terephthalate	1 PET or PETE	Clear, lightweight but strong and heat-resistant. Good barrier to gas, moisture, alcohol, and solvents.	• Water bottles • Food jars • Ovenproof film
High-density polyethylene	2 HDPE	Tough; can be stretched without breaking, and easy to process. Resistant to moisture and solvents.	• Milk containers • Wheelie bins • Juice bottles
Polyvinyl chloride	3 PVC	Strong; resistant to chemicals and oil. Rigid PVC is used in construction; flexible PVC in inflatables.	• Pipes • Toys and inflatables • Flooring
Low-density polyethylene	4 LDPE	Flexible and tough, can withstand high temperatures. Good resistance to chemicals. Easy to process.	• Plastic bags • Snap-on lids • Six-pack rings
Polypropylene	5 PP	Tough, flexible, and long lasting. High melting point. Resistant to fats and solvents.	• Hinges on flip-top lids • Plastic medicine bottles • Concrete additives
Polystyrene	6 PS	Can be solid or foamed. Good for insulation and easy to shape, but slow to biodegrade.	• Disposable foam cups • Plastic cutlery • Packaging
Miscellaneous	7 Miscellaneous	Other plastics such as acrylic, nylon, polylactic acid, and plastic multi-layer combinations.	• Baby bottles • Safety glasses • "Ink" in 3D printers

Corrosive power

Strong acids and alkalis can cause serious burns to skin. Very strong acids and alkalis can burn through metal, and some can even dissolve glass. While dangerous, their corrosive power can be useful, for instance, for etching glass or cleaning metals.

Acids and bases

Chemical opposites, acids and bases react when they are mixed together, neutralizing one another. Bases that are soluble in water are called alkalis. All alkalis are bases, but not all bases are alkalis.

Bases and acids can be weak or strong. Many ingredients in food contain weak acids (vinegar, for instance) or alkalis (eggs), while strong acids and alkalis are used in cleaning products and industrial processes. Strong acids and alkalis break apart entirely when dissolved in water, whereas weak acids and alkalis do not.

Is it an acid or a base?

The acidity of a substance is measured by its number of hydrogen ions – its "power of hydrogen" or pH. Water, with a pH of 7, is a neutral substance. A substance with a pH lower than 7 is acidic; one with a pH above 7 is alkaline. Each interval on the scale represents a tenfold increase in either alkalinity or acidity. For instance, milk, with a pH of 6, is ten times more acidic than water, which has a pH of 7. Meanwhile, seawater, with a pH of 8, is ten times more alkaline than pure water.

Hydrogen ions (H⁺)

determine whether a solution is an acid or an alkali. Acids are H⁺ donors while alkalis are H⁺ acceptors.

The pH scale

Running from 0 to 14, the pH scale is related to the concentration of hydrogen ions (H⁺). A pH of 7 is neutral. A pH of 1 indicates a high concentration of hydrogen ions (acidic). A pH of 14 shows a low concentration (alkaline).

The litmus test

A version of the litmus test has been used for hundreds of years to tell whether a solution is acidic or alkaline. Red litmus paper turns blue when dipped into an alkali. Blue litmus paper turns red when dipped into an acid.

Red colouring indicates acid.

Blue colouring indicates alkali.

Stomach acid is corrosive.

Vinegar is a weak acid.

Apples are acidic.

Milk is slightly acidic.

Pure water is neutral.

Seawater is slightly alkaline.

Baking soda has a pH of 9.

Drain cleaners are strong alkalis.

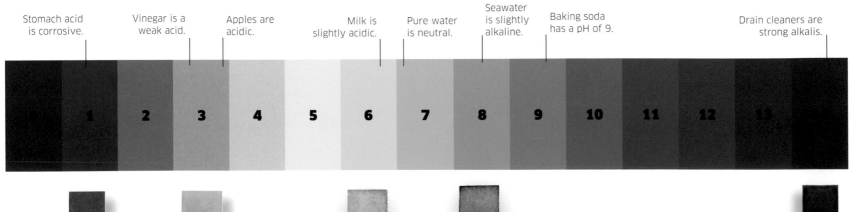

1 2 3 4 5 6 7 8 9 10 11 12 13

The universal indicator test

Indicator paper contains several different chemicals that react, turning a range of colours in response to different pH values. Dipping indicator paper into an unknown solution reveals its pH.

Gardeners use coffee grounds to lower the pH of soil around acid-loving plants such as roses.

Soapy water is strongly alkaline.

Stomach acid is almost as corrosive as battery acid, but our stomachs produce a mucus that protects us from damage.

47

It's all about the ions

The difference between an acid and an alkali comes down to their proportion of positively charged particles called hydrogen ions (H⁺). When an acidic compound is dissolved in water, it breaks up, releasing H⁺ ions: it has an increased proportion of positively charged ions. When an alkaline compound dissolves in water it releases negatively charged particles called hydroxide ions (OH⁻). Acids are called H⁺ donors; alkalis are called H⁺ acceptors.

Acid
There are more positively charged H⁺ ions than negatively charged OH⁻ ions in an acid.

Neutral
A neutral solution contains equal numbers of positive H⁺ and negative OH⁻ ions.

Base
There are more negatively charged OH⁻ ions than positively charged H⁺ ions in an alkali.

Mixing acids and bases

The reaction between an acid and an alkali produces water and a salt. It is called a neutralization reaction. The H⁺ ions in the acid react with the OH⁻ ions in the alkali, resulting in a substance that is neither acid nor alkali. Different acids and alkalis produce different salts when they react.

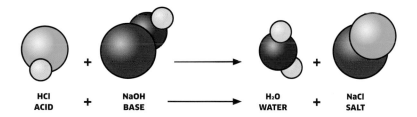

$$HCl\ (ACID) + NaOH\ (BASE) \rightarrow H_2O\ (WATER) + NaCl\ (SALT)$$

Neutralization formula

When hydrochloric acid (HCl) reacts with the alkali sodium hydroxide (NaOH), they produce a neutral solution that consists of water (H₂O) and a well-known salt – sodium chloride (NaCl), or common salt.

Acids and bases in agriculture

Farmers monitor soil pH levels carefully. Soils are naturally acidic or alkaline, and different crops prefer a higher or lower pH. Farmers can reduce the soil pH by adding certain fertilizers, or raise the soil pH with alkalis, such as lime (calcium hydroxide).

Kitchen chemistry

The kitchen is a great place to see acids and alkalis in action. Weak acids – found in lemon juice and vinegar – can preserve or improve the flavour of food. When baking, we use weak alkalis present in bicarbonate of soda to help cakes to rise. Strong acids and alkalis are key ingredients in a range of cleaning products. They are so powerful that protective gloves must be worn when using them.

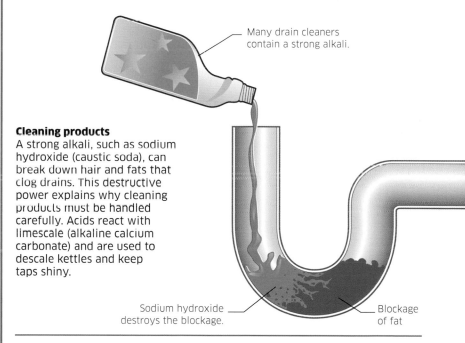

Many drain cleaners contain a strong alkali.

Cleaning products

A strong alkali, such as sodium hydroxide (caustic soda), can break down hair and fats that clog drains. This destructive power explains why cleaning products must be handled carefully. Acids react with limescale (alkaline calcium carbonate) and are used to descale kettles and keep taps shiny.

Sodium hydroxide destroys the blockage.

Blockage of fat

Bubbles made by carbon dioxide

Baking powder

Added to flour to help cakes rise, baking powder contains an acid and an alkali, which react together when a liquid and heat are added. The reaction produces bubbles of carbon dioxide that push the cake mixture upwards.

Crystal forest

If you dip a piece of pure metal into a solution in which another metal is dissolved, something quite magical may happen.

These delicate crystals have formed on a piece of zinc placed in a solution of lead nitrate. The magic is in fact a chemical reaction known as metal displacement, seen here in a photograph taken through a microscope. The more reactive metal (zinc) displaces the less reactive metal (lead) from its nitrate compound, so instead of lead nitrate and zinc, we end up with pure lead and a solution of zinc and nitrate ions. The lead atoms join together in regular patterns, forming crystals of pure lead.

Combustion

Combustion is the reaction between a fuel – such as wood, natural gas, or oil – and oxygen. The combustion reaction releases energy in the form of heat and light. Fuel needs a trigger (a match or a spark) before combustion can start.

Combustion is at work in bonfires, fireworks, and when we light a candle. But more than just a spectacle, it is essential to the way we live. Most of the world's power stations generate electricity using the combustion of fossil fuels such as coal, oil, and gas. Most cars, lorries, boats, and planes are driven by engines powered by combustion. Scientists are working hard to create alternatives to what is now understood to be a potentially wasteful and harmful source of energy. But for now we all rely on it to keep warm and to get where we need.

Campfire chemistry

Dry wood contains cellulose (made of the elements carbon, hydrogen, and oxygen). It burns well in oxygen, which makes up about one fifth of air.

Carbon dioxide
Carbon dioxide (CO_2) is produced when wood burns. Known as a greenhouse gas, it contributes to global warming if there is too much of it in the atmosphere.

Water vapour
The combustion of cellulose, which makes up about half the dry mass of wood, produces water (H_2O) as well as carbon dioxide (CO_2). In the heat of a fire, the water evaporates as steam.

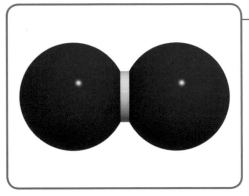

Oxygen
For combustion to work, there needs to be a good supply of the element oxygen. Oxygen in the air exists as molecules made up of two oxygen atoms, with the chemical formula O_2.

A balanced reaction

During combustion, substances known as reactants are transformed into new substances called products. The reaction rearranges the atoms of the reactants. They swap places, but the number of each is the same. Energy (heat and light) is released when the bonds that hold the initial molecules together are broken and new ones are formed.

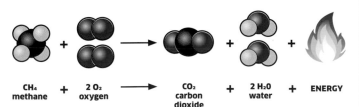

| CH_4 methane | + | $2 O_2$ oxygen | | CO_2 carbon dioxide | + | $2 H_2O$ water | + | ENERGY |

Methane combustion

Above is the reaction formula for the combustion of methane (natural gas). The number of carbon (C), hydrogen (H), and oxygen (O) atoms is the same on each side of the arrow, but the substances they make up have changed.

Early man first learned to make fire around one million years ago.

1777 The year French chemist **Antoine Lavoisier** proved that **oxygen** is involved **in combustion**.

51

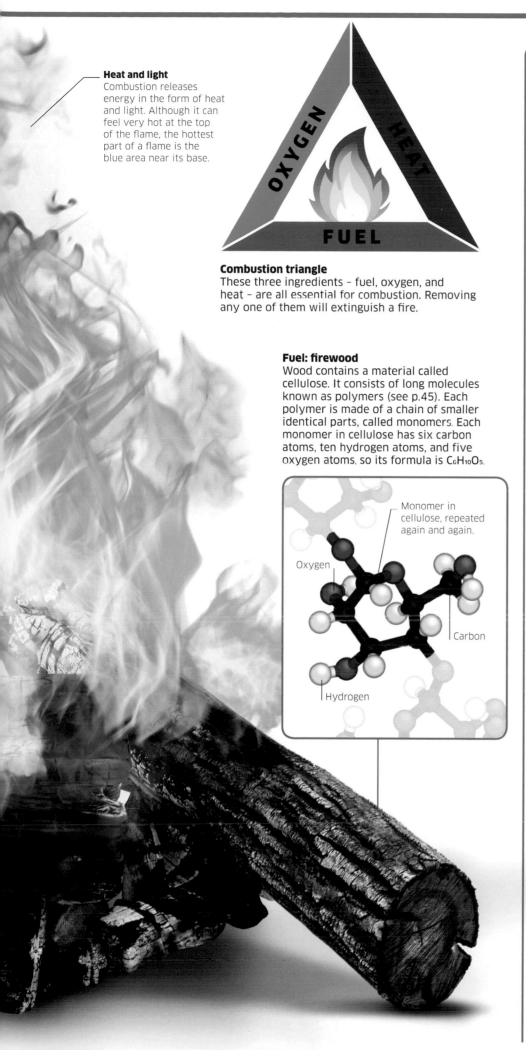

Heat and light
Combustion releases energy in the form of heat and light. Although it can feel very hot at the top of the flame, the hottest part of a flame is the blue area near its base.

OXYGEN HEAT FUEL

Combustion triangle
These three ingredients – fuel, oxygen, and heat – are all essential for combustion. Removing any one of them will extinguish a fire.

Fuel: firewood
Wood contains a material called cellulose. It consists of long molecules known as polymers (see p.45). Each polymer is made of a chain of smaller identical parts, called monomers. Each monomer in cellulose has six carbon atoms, ten hydrogen atoms, and five oxygen atoms, so its formula is $C_6H_{10}O_5$.

Monomer in cellulose, repeated again and again.

Oxygen

Carbon

Hydrogen

Fuel efficiency and the environment
Different fuels release different amounts of energy. They also produce different amounts of carbon dioxide when they burn. Wood is least efficient, and produces the most carbon dioxide, which makes it the least environmentally friendly fuel.

Energy values of different fuels
- Energy content (kJ per gram of fuel)
- Quantity (mg) of carbon dioxide released per kJ of energy

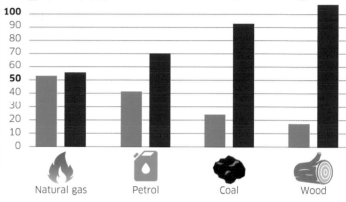

100 90 80 70 60 50 40 30 20 10 0

Natural gas   Petrol   Coal   Wood

Fireworks
Fireworks shoot up in the air and explode into colourful displays thanks to combustion. The fuel used is charcoal, mixed with oxidisers (compounds providing oxygen) and other agents. The colours come from different metal salts.

Sodium salts make yellow stars.

Green comes from barium salts.

Copper salts mixed with red strontium salts make purple.

Copper salts make blue stars.

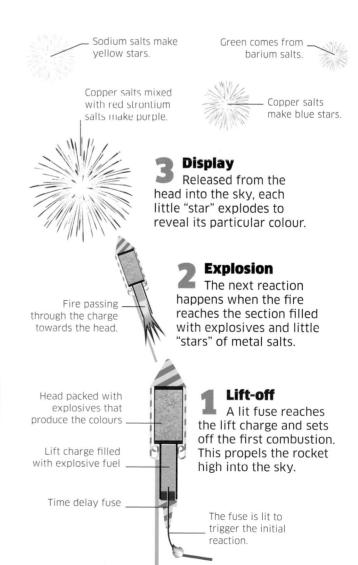

3 Display
Released from the head into the sky, each little "star" explodes to reveal its particular colour.

2 Explosion
The next reaction happens when the fire reaches the section filled with explosives and little "stars" of metal salts.

Fire passing through the charge towards the head.

1 Lift-off
A lit fuse reaches the lift charge and sets off the first combustion. This propels the rocket high into the sky.

Head packed with explosives that produce the colours

Lift charge filled with explosive fuel

Time delay fuse

The fuse is lit to trigger the initial reaction.

Electrochemistry

Electricity and chemical reactions are closely linked, and together fall under the heading electrochemistry. Electrochemistry is the study of chemical processes that cause electrons to move.

An electric current is a steady flow of electrons, the tiny negative particles that whizz around in the shells of atoms. Electrons can flow in response either to a chemical reaction taking place inside a battery, or to a current delivered by the mains electricity supply.

Electricity is key to electrolysis. This process is used in industries to extract pure elements from ionic compounds (see p.44) that have been dissolved in a liquid known as an electrolyte. Electrolysis can also be used to purify metals, and a similar process can be used to plate (cover) objects with a metal. The result depends on the choice of material of the electrodes and, in particular, the exact contents of the electrolyte.

Ions and redox reactions

Chemical reactions where electrons are transferred between atoms are called oxidation-reduction (redox) reactions. Atoms that have lost or gained electrons become ions, and are electrically charged. Atoms that gain electrons become negative ions (anions). Atoms that lose electrons become positive ions (cations). These play an important role in electrolysis.

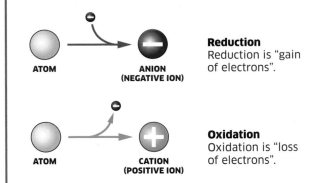

ATOM → **ANION (NEGATIVE ION)**

Reduction
Reduction is "gain of electrons".

ATOM → **CATION (POSITIVE ION)**

Oxidation
Oxidation is "loss of electrons".

Electrolysis

Ionic compounds contain positive and negative ions. They can be separated using electricity, by a process called electrolysis. If electricity passes through an electrolyte (an ionic compound that has been dissolved in water), the negative ions in the electrolyte will flow towards the positive electrode and the positive ions will flow towards the negative electrode. The products created in the process will depend on what is in the electrolyte. This diagram shows how water (H_2O) can be split back into its original pure elements, oxygen and hydrogen. The two gases can be trapped and collected as they bubble up along the electrodes.

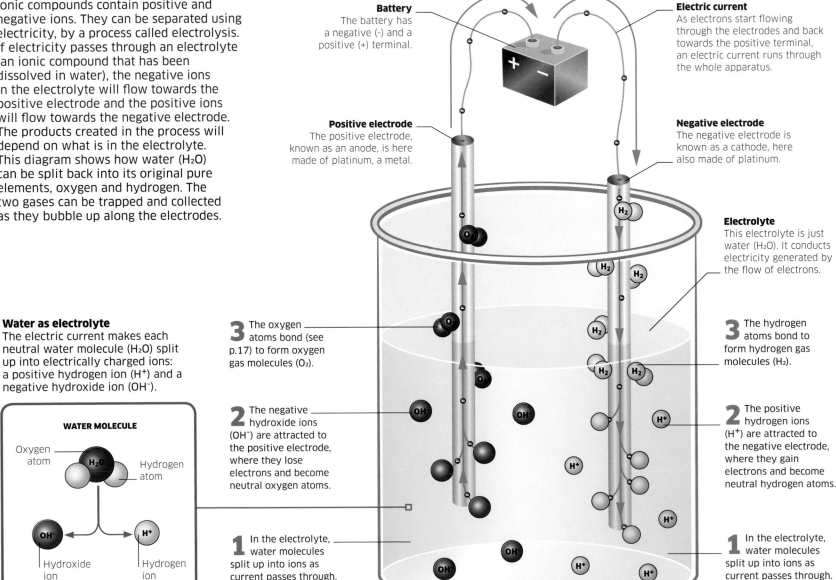

Battery
The battery has a negative (-) and a positive (+) terminal.

Electric current
As electrons start flowing through the electrodes and back towards the positive terminal, an electric current runs through the whole apparatus.

Positive electrode
The positive electrode, known as an anode, is here made of platinum, a metal.

Negative electrode
The negative electrode is known as a cathode, here also made of platinum.

Electrolyte
This electrolyte is just water (H_2O). It conducts electricity generated by the flow of electrons.

Water as electrolyte
The electric current makes each neutral water molecule (H_2O) split up into electrically charged ions: a positive hydrogen ion (H^+) and a negative hydroxide ion (OH^-).

WATER MOLECULE

Oxygen atom — H_2O — Hydrogen atom

OH^- Hydroxide ion H^+ Hydrogen ion

3 The oxygen atoms bond (see p.17) to form oxygen gas molecules (O_2).

2 The negative hydroxide ions (OH^-) are attracted to the positive electrode, where they lose electrons and become neutral oxygen atoms.

1 In the electrolyte, water molecules split up into ions as current passes through.

3 The hydrogen atoms bond to form hydrogen gas molecules (H_2).

2 The positive hydrogen ions (H^+) are attracted to the negative electrode, where they gain electrons and become neutral hydrogen atoms.

1 In the electrolyte, water molecules split up into ions as current passes through.

1807 The year British chemist **Sir Humphry Davy** discovered the **elements** potassium and sodium **through electrolysis**.

The **human body contain electrolytes** that regulate nerve and muscle functions, which rely on a **weak electric current**.

53

Electroplating

Similar to electrolysis, electroplating is a process that coats a cheaper metal with a more expensive metal, such as silver. To turn a cheap metal spoon into a silver-plated spoon, the cheap metal spoon is used as a cathode (negative electrode) and a silver bar is used as the anode (positive electrode). These two electrodes are bathed in an electrolyte that contains a solution of the expensive metal, in this case silver nitrate solution.

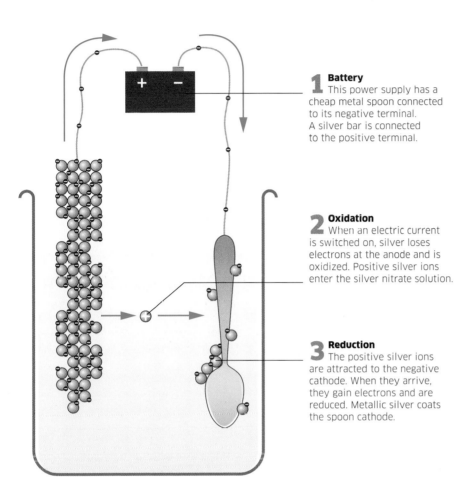

1 Battery
This power supply has a cheap metal spoon connected to its negative terminal. A silver bar is connected to the positive terminal.

2 Oxidation
When an electric current is switched on, silver loses electrons at the anode and is oxidized. Positive silver ions enter the silver nitrate solution.

3 Reduction
The positive silver ions are attracted to the negative cathode. When they arrive, they gain electrons and are reduced. Metallic silver coats the spoon cathode.

Grey tarnish showing that oxidation has taken place.

Oxidation in air
Silver oxidizes when exposed to air, so silver plated items eventually lose their shine as a grey tarnish forms on the surface. Polishing removes the tarnish but the plating might be damaged.

Galvanizing
Steel or iron can be prevented from rusting (a form of oxidation) by coating them in the metal zinc, a process called galvanizing. These nails have been galvanized.

Purifying metals

The copper that is extracted from copper ore is not pure enough to become electrical wiring. It has to be purified by electrolysis. Impure copper acts as the anode, and pure copper as the cathode. These electrodes lie in a solution of copper sulfate.

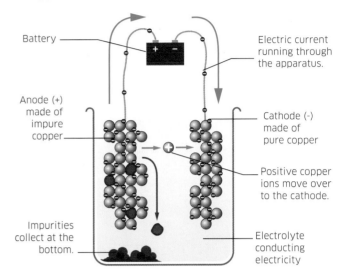

Battery

Electric current running through the apparatus.

Anode (+) made of impure copper

Cathode (-) made of pure copper

Positive copper ions move over to the cathode.

Impurities collect at the bottom.

Electrolyte conducting electricity

Electrorefining

Pure copper is used to make electrical wiring and components. Here you can see copper purification, called electrorefining, being carried out on a massive scale in a factory, in the process described above.

Electrochemistry in batteries

Batteries turn chemical energy into electrical energy (see p.92). This is the opposite of electrolysis, which turns electrical energy into chemical energy. In a battery, it is the anode that is negative and the cathode that is positive. The reaction at the anode is still oxidation and at the cathode it is still reduction.

Cathode (manganese dioxide), mixed with alkaline electrolyte.

Anode (zinc and carbon paste)

Brass pin conducts electrons to the negative terminal.

Positive terminal (steel cap)

Negative terminal

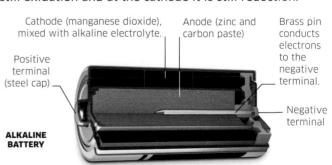

ALKALINE BATTERY

Hot metal

Ignite a mixture of chemicals called thermite and you'll need to stand back! These chemicals react together very quickly, producing enormous amounts of heat.

A thermite reaction is a spectacular display, but also serves a practical purpose: it is used to extract molten iron from iron oxide for welding. It takes a lot of heat to start the reaction, which then releases enough heat to melt the iron. The process most commonly uses a mixture of iron oxide and aluminium. A slim ribbon of magnesium is inserted into the mixture as a fuse. When ignited, it starts the reaction, breaking the bonds between the iron and oxygen atoms. Aluminium then bonds with the released oxygen, producing more heat. This in turn breaks more bonds and melts the leftover iron.

58 matter ∘ **NATURAL MATERIALS**

600 The **number of trees** needed to build a medieval **warship**.

Natural materials

Early humans learned to use the materials they found around them to make tools, clothes, and homes. Many natural materials are still used in the same way, while others are combined to make new ones.

Some natural materials come from plants (for example wood, cotton, and rubber), others from animals (silk and wool), or from Earth's crust (clay and metals). Their natural properties – bendy or rigid, strong or weak, absorbent or waterproof – have been put to good use by humans for millions of years. People have also learnt to adjust these properties to suit their needs. Soft plant fibres and animal wool are spun into longer, stronger fibres. Animal skins are treated to make leather to wear. Skins were also used to make parchment to write on; now we use paper made from wood. Metals are mixed to make stronger materials called alloys (see pp.62–63).

Materials from plants

Plant materials have played a key role in humans' success as a species. Wood has provided shelters, tools, and transport, while cotton and flax (a plant used to make linen) have clothed people for thousands of years. Plant materials can be flexible or rigid, heavy or light, depending on the particular combination of three substances in their cell walls: lignin, cellulose, and hemicellulose.

Latex and rubber
Today, a lot of rubber is synthetic, but natural rubber comes from latex, a fluid that can be tapped from certain types of trees. It contains a polymer that makes it elastic.

Cotton
Fluffy cotton, consisting mainly of cellulose, protects the cotton plant's seeds. It is picked and spun into yarn or thread. The texture of cotton fabrics vary depending on how they are woven.

Wood
Different types of wood have different properties, including colour, texture, weight, and hardness, making them suitable for different things. Wood pulp is used to make paper. A lot of wood is harvested from wood plantations.

Materials from animals

Animals, from insects to mammals, are a rich source of materials. The skin of pigs, goats, and cows can be treated and turned into leather. Caterpillars called silkworms spin themselves cocoons that can be unravelled into fine silk threads. Sheep grow thick, waterproof hair that can be cut off, or shorn, and spun into wool thread used for knitting or woven into fabrics.

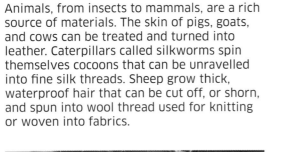

The silk used for these bright scarves has been dyed. Natural silk is pale in colour, and its tone depends on what the silkworms are eating.

Silk
Silkworms and their moth parents have been farmed for more than 5,000 years. A cocoon can produce up to 900 m (2,950 ft) of silk thread that can be made into beautiful fabrics.

Different breeds of sheep produce different types of wool.

Wool yarn
Wool is washed, then spun into long fibres, and dyed.

Wool
Sheep have been bred for their wool for more than 6,000 years. An average sheep produces wool for about eight jumpers a year, or 60 pairs of socks. Today, wool is often mixed with acrylic fibres.

Bamboo, a fast-growing, **tree-like grass**, can be **turned into a fabric** that is **soft, breathable, and absorbs sweat**, making it good for sportswear.

Glass was first made in Ancient Egypt and Mesopotamia in **around 2,000 BCE**.

59

Keeping it natural
Natural materials, such as rubber, cotton, and different types of wood, are used in a wide range of everyday items, such as the ones seen here.

Vulcanized tyres
Adding sulfur to natural rubber, a process called vulcanization, increases its durability.

Elastic, not plastic
Washing-up gloves are often made of flexible latex.

Thin but strong
Cellulose polymer chains line up together to give cotton thread its strength.

Absorbent cotton
Cotton is great for towels and cotton buds as it is soft and can absorb up to 27 times its weight in water.

Cotton buds

Steady support
Lignin is the substance that holds cellulose and hemicellulose fibres together and makes wood stiff and strong – useful properties for ladders.

Curved wood
Some woods, such as maple and spruce, can be bent into shape using steam. They are good for making violins and other string instruments.

Materials from Earth's crust
Earth materials range from sand, clay, and rocks to minerals and metals. Materials from the earth have always been important for building. If you look at buildings, you can usually see what materials lie underground in the area – flint or slate, sandstone, limestone, marble, or clay. These materials are also essential for practical and decorative cookware, crockery, and utensils.

Clay and clay products
Clay, a mixture of the minerals silicon dioxide and aluminium oxide, has many uses. To make bricks, natural clay is mixed with water and pressed into shape before being dried. It is then baked at very hot temperatures to make it waterproof. Pottery is made in a similar way, but with clay of finer particles.

Earthenware pottery is fired at temperatures of around 1,000°C (1,830°F).

Sand and glass
Glass is made from sand. It is usually the sand common in deserts, which consists of the mineral silica. Beach sand often has traces of other substances, so makes less clear glass. Carefully chosen additives colour the glass. The ingredients are melted together at 1,500°C (2,732°F) before being shaped into window panes, drinking glasses, or bottles.

Spectacle lenses used to be made of pure glass. Today they are often plastic.

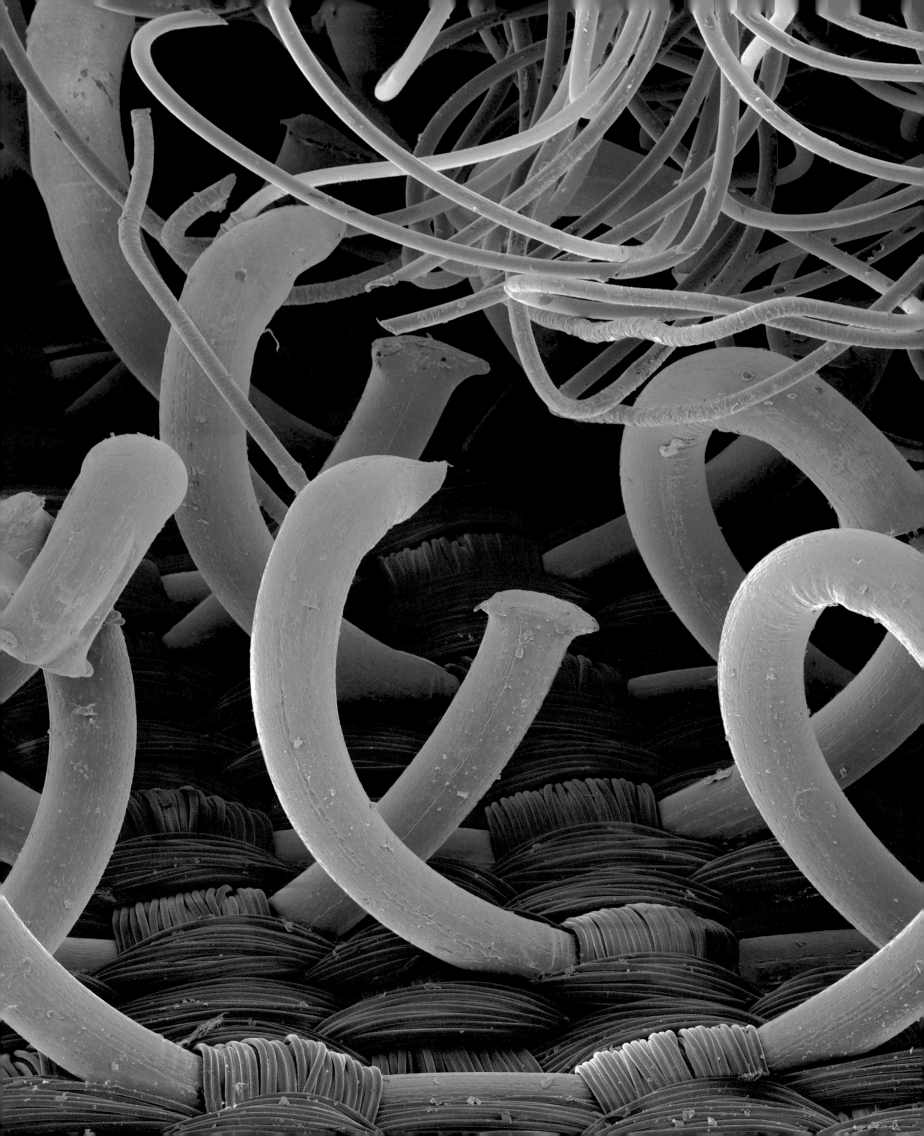

Hook and loop

Most people are familiar with Velcro®, a quick and easy fastener on clothing, shoes, and bags. This is what it looks like close-up.

This false-coloured image, captured by an electron microscope, shows the small, soft loops (blue) that catch in the sturdy hooks (green) when the two strips are pressed together. Velcro is made from nylon or polyester. It was invented by the Swiss engineer George de Mestral in 1941, after he noticed that hooked burdock seeds stuck to his dog's fur and to his own clothing.

62 matter ∘ ALLOYS

1874 The year that the **Eads Bridge**, the **world's first bridge** all **made of steel**, opened for traffic across **the Missouri river, USA**.

Alloys

An alloy is a mixture of at least two different elements, at least one of which is a metal. Alloys are used to make many things, including car and aeroplane parts, musical instruments, jewellery, and medical implants.

In many alloys, all the elements are metal. But some alloys contain non-metals, such as carbon. The ingredients of an alloy are carefully chosen for the properties they bring to the alloy, whether to make it stronger, more flexible, or rust-resistant. All alloys have metallic properties and are good electrical conductors, and all have many advantages over pure metals.

Early alloys

The first man-made alloy was bronze. It was developed around 5,000 years ago by smelting (heating) copper and tin together. This was the start of the Bronze Age, a period in which this new, strong alloy revolutionized the making of tools and weapons. Some thousand years later, people learned to make brass from copper and zinc.

Bronze weapons
Bronze can be hammered thin, stretched, and moulded. These objects, made in Mesopotamia around 2000 BCE, were designed to fit on a mace (a club-like weapon).

Atomic arrangements

It is how the atoms are arranged in a material that decides how it behaves in different conditions. Atoms of pure metals are regularly arranged, but in alloys this arrangement is disrupted. The atoms of the main component of an alloy may be of a similar size, or much bigger, than those of the added one. They can be arranged in several ways.

Identical atoms of pure metals

Pure metals
The atoms in a pure metal such as gold (left) are neatly arranged. Under pressure, they will slide over one another, causing cracking.

Zinc atoms replace copper atoms in a brass alloy used for trumpets.

Substitutional alloys
Atoms of the added component take up almost the same space as atoms of the main one. This distorts the structure and makes it stronger.

Tiny carbon atoms sit between large iron atoms, making steel very strong.

Interstitial alloys
These alloys, such as steel used for bridges, are strong: smaller atoms fill the gaps between larger ones, preventing cracking or movement.

Interstitial carbon atoms and substitutional nickel or chromium atoms make stainless steel strong as well as non-rusting.

Combination alloys
Some alloys have a combination of atom arrangements to improve their properties. An example is stainless steel, used in cutlery.

Alloys in coins

Coins used to be made of gold and silver, but these metals are too expensive and not hard-wearing enough for modern use. Several different alloys are used for coins today. They are selected for their cost, hardness, colour, density, resistance to corrosion, and for being recyclable.

EU €2-coin
Outer ring: copper (75%), nickel (25%). Centre: copper (75%), zinc (20%), nickel (5%).

British £1-coin
Outer ring: copper (76%), zinc (20%), nickel (4%). Inner ring: copper (75%), nickel (25%).

Egyptian £1-coin
Outer ring: steel (94%), copper (2%), nickel plating (4%). Inner ring: steel (94%), nickel (2%), copper plating (4%).

Australian $1-coin
Copper (92%), nickel (2%), aluminium (6%).

Sterling silver, used in most silver **jewellery**, is in fact an **alloy**, containing **7.5 per cent copper**.

Mercury makes up about half of **amalgam, an alloy** sometimes used as **dental filling**; the rest is **silver, copper, and tin**.

63

Clever alloys

All alloys are developed to be an improvement on the individual metals from which they were made. Some alloys are an extreme improvement. Superalloys, for example, have incredible mechanical strength, resistance to corrosion, and can withstand extreme heat and pressure. These properties makes them very useful in aerospace engineering, as well as in the chemical industries. Memory alloys, or smart alloys, often containing nickel and titanium, "remember" their original shape.

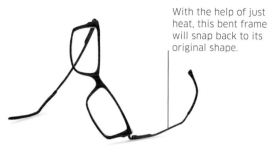

With the help of just heat, this bent frame will snap back to its original shape.

Memory alloys
An object made from a memory alloy can return to its original shape if it has been bent. Simply applying heat restores the alloy to the shape it was in.

Superalloy used in jet engine

Superalloys
These high-performance alloys hold their shape in temperatures close to their high boiling points of around 1,000°C (1,832°F).

Spanish piece-of-eight
These legendary Spanish coin were made of silver. From the 15th to the 19th centuries, they were used throughout the vast Spanish Empire, and in other countries, too.

Japanese 50-yen coin
Copper (75%) and nickel (25%).

US dime (10-cent) coin
Copper (91.67%) and nickel (8.33%).

Swedish 10-krona coin
An alloy known as "Nordic gold", also used in euro cents: copper (89%), aluminium (5%), zinc (5%), tin (1%).

Aluminium alloys

The metal aluminium is lightweight, resistant to corrosion, and has a high electrical conductivity. It is useful on a small scale (as foil, for example) but, because it is soft, it needs to be alloyed with other elements to be strong enough to build things. Aluminium alloys are often used in car bodies and bicycle frames.

Light, rust-proof frame made of an aluminium alloy

Steel

Iron, a pure metal, has been used since the Iron Age, some 3,000 years ago. But although it is very strong, iron is also brittle. There were some early iron alloys, but the strongest one, steel, came into common use during the Industrial Revolution in the 19th century. There are two ways of making steel: it can be produced from molten "pig iron" (from iron ore) and scrap metal in a process called basic oxygen steelmaking (BOS), or from cold scrap metal in the electric arc furnace (EAF) process. Impurities, such as too much carbon, are removed, and elements such as manganese and nickel are added to produce different grades of steel. The molten steel is then shaped into bars or sheets ready to make into various products.

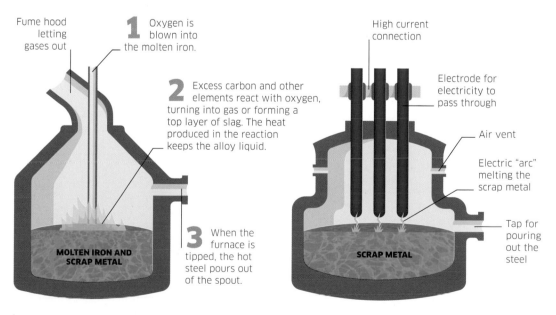

Fume hood letting gases out

1 Oxygen is blown into the molten iron.

2 Excess carbon and other elements react with oxygen, turning into gas or forming a top layer of slag. The heat produced in the reaction keeps the alloy liquid.

3 When the furnace is tipped, the hot steel pours out of the spout.

MOLTEN IRON AND SCRAP METAL

High current connection

Electrode for electricity to pass through

Air vent

Electric "arc" melting the scrap metal

Tap for pouring out the steel

SCRAP METAL

Basic oxygen steelmaking (BOS)
Oxygen is blown through molten "pig iron" and scrap metal to reduce its carbon content and other impurities. Then alloying elements are added, turning the molten metal into steel.

Electric arc furnace (EAF)
Cold scrap metal is loaded into the furnace. An electric current forms an "arc" (a continuous spark), which melts the metal. The final grade of steel is determined by adding alloying elements.

64   matter ○ **MATERIALS TECHNOLOGY**

9 billion tonnes of **plastics** have been **made** since the 1950s.

Materials technology

Synthetic materials are born in laboratories. Using their knowledge of elements and compounds, chemists can develop new materials with unique properties, created for specific tasks.

Materials created artificially perform different functions depending on their chemistry – the arrangements of their atoms or molecules, and how they react. Research constantly brings new materials to meet new challenges, ranging from synthetic textiles and biodegradeable plastics to the vast range of high-performance materials that make up a racing car.

Fuel tank
Combining bullet-proof Kevlar and flexible rubber keeps the tank light, strong, and less likely to crack on impact.

Brakes
Adding carbon fibre to brake discs keeps them light and able to resist temperatures of up to 1,200°C (2,192°F).

Exhaust
This is formed from a 1 mm (0.04 inch) thick heat-resistant steel alloy, first made for the aerospace industry.

Engine
Precise regulations decide which materials can be used for the many parts of a Formula One engine – no composites are allowed.

Racing car
Formula One cars rely on materials that can withstand extreme heat and pressure. The structure must be rigid in some parts and flexible in others; some parts are heavy while some have to be light. The driver, also exposed to heat and pressure, and speeds over 320 km/h (200 mph), relies on synthetic materials to keep safe. Their clothing is made with layers of Nomex®, a fire-resistant polyamide (a type of plastic) used for fire and space suits. Kevlar®, similar to Nomex but so strong it is bullet proof, is used to reinforce various car parts, as well as the driver's helmet.

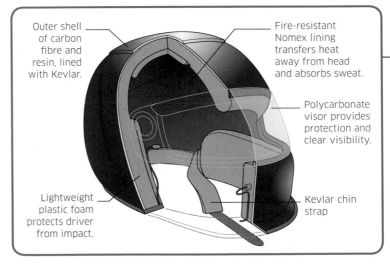

Outer shell of carbon fibre and resin, lined with Kevlar.

Fire-resistant Nomex lining transfers heat away from head and absorbs sweat.

Polycarbonate visor provides protection and clear visibility.

Lightweight plastic foam protects driver from impact.

Kevlar chin strap

Helmet anatomy
The driver is subjected to extreme G-forces when braking and cornering. This puts great strain on their neck. To help keep their head up, their helmet must be as light as possible. Highly specialized materials are used for the helmet, which needs to be comfortable and light, yet strong and able to absorb impacts and resist penetration, in case of an accident.

The **lightest man-made** solid is **aerogel**, which is both **fireproof** and insulating.

A **waterproof superglue** that one day could be **used to heal wounds** is based on the **sticky slime** that keeps **mussels** stuck to rocks.

65

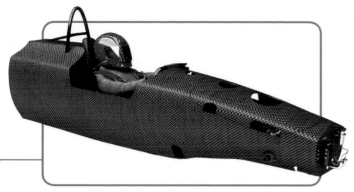

Survival cell

The monocoque, or survival cell, surrounds the cockpit where the driver sits. It is made of a strong, stiff carbon-fibre composite that can absorb the full energy of an impact without being damaged. Carbon fibre is much lighter than steel or aluminium, helping the car go faster and use less fuel.

Mimicking nature

Many synthetic materials were invented to replace natural materials that were too hard, or too expensive, to extract or harvest. For example, nylon was invented to replace silk in fabrics, and polyester fleece can be used instead of wool. Ever advancing technology makes it possible to imitate some amazing materials, such as spider silk, which is tougher than Kevlar, stronger than steel, yet super flexible.

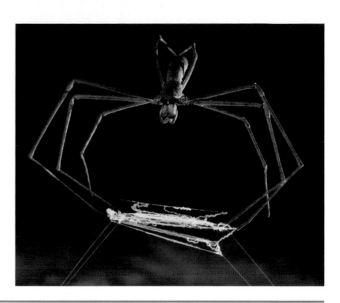

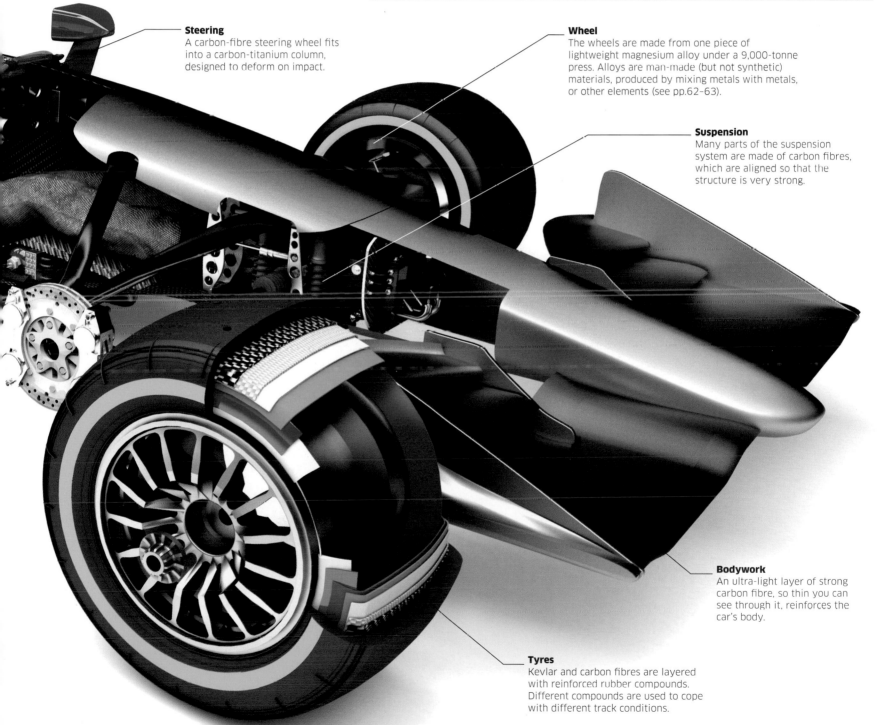

Steering
A carbon-fibre steering wheel fits into a carbon-titanium column, designed to deform on impact.

Wheel
The wheels are made from one piece of lightweight magnesium alloy under a 9,000-tonne press. Alloys are man-made (but not synthetic) materials, produced by mixing metals with metals, or other elements (see pp.62–63).

Suspension
Many parts of the suspension system are made of carbon fibres, which are aligned so that the structure is very strong.

Bodywork
An ultra-light layer of strong carbon fibre, so thin you can see through it, reinforces the car's body.

Tyres
Kevlar and carbon fibres are layered with reinforced rubber compounds. Different compounds are used to cope with different track conditions.

ENERGY AND FORCES

Energy and forces are essential concepts in science, for nothing can happen without them. Forces change the motion of an object, and energy is behind everything that changes – from a flower opening to an exploding bomb. The amount of energy in the Universe is fixed; it cannot be created or destroyed.

The modern age
Scientific discovery and technology go hand in hand as astronomers and physicists use computer science and particle accelerators to expand our knowledge of the Universe.

Relativity
Einstein's General Theory of Relativity explains that what we perceive as gravity is an effect of the curvature of space and time.

MODEL OF THE EXPANDING UNIVERSE

Big Bang theory
Belgian priest and physicist Georges Lemaître comes up with the theory of an ever-expanding universe that began with the Big Bang – the source of all energy and forces.

20TH CENTURY

1886

1916

1927

Radio waves
German physicist Heinrich Hertz proves that electromagnetic waves exist.

1848

THERMOMETER

Combustion engine
German engineer Nikolaus Otto develops the internal combustion engine. It uses understanding gained over two hundred years of how the temperature, volume, and pressure of gases relate.

Absolute zero
Scottish scientist Lord Kelvin calculates the lowest possible temperature, at which particles almost cease to vibrate, as −273°C (−460°F), calling it absolute zero.

RADIO MAST

OTTO'S ENGINE

1876

Discovering energy and forces

People have been asking questions about how the world around them works, and using science to find answers for them, over thousands of years.

From the forces that keep a ship afloat and the magnetism that helps sailors to navigate the oceans with a compass, to the atoms and sub-atomic particles that make up our world and the vast expanses of space, people through history have learned about the Universe by observation and experiment. In ancient and medieval times, as the tools available to study the world were limited, so was knowledge of science. The modern scientific method is based on experiments, which are used to test hypotheses (unproven ideas). Observed results modify hypotheses, improving our understanding of science.

Gravity
English scientist Isaac Newton (left) explains how gravity works after an apple falls on his head.

18TH CENTURY

ISAAC NEWTON

1687

1678

Wave theory of light
Dutch scientist Christiaan Huygens announces his theory that light travels in waves. This is contested by Newton's idea that light is made of particles.

LIGHT AS A WAVE

Ancient and medieval ideas
The ancient Greeks and Romans used debate to help them understand the Universe, while Arab and Chinese scholars studied mathematics and natural phenomena such as rainbows and eclipses.

Buoyancy
The Greek thinker Archimedes realizes the force pushing upwards on an object in water is equal to the weight of water displaced.

Magnetic compass
The Chinese create primitive compasses with lodestone, a naturally occurring magnet.

Light vision
Arab scholar Alhazen suggests that light is emitted from objects into the eye, not the reverse.

BEFORE 1500

240 BCE

200 BCE

1011 CE

BOMBE CODE-BREAKING MACHINE

Computer science
British code-breaker Alan Turing develops the first programmable computer, laying the foundations of modern computer science.

Nuclear energy
Italian-American physicist Enrico Fermi leads a US team that builds the world's first nuclear fission reactor. In 1945, the first atomic bomb is dropped on Hiroshima.

NUCLEAR EXPLOSION

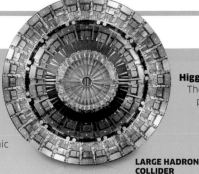

Higgs boson
The Higgs boson particle is identified, confirming the Standard Model of particle physics developed in the 1970s.

LARGE HADRON COLLIDER

1890-PRESENT

1936   1942   2012

1831   1799

Energy conservation
German physicist Hermann von Helmhotz states that energy cannot be created or destroyed, it can only change its form.

PERPETUAL MOTION MACHINE

Electromagnetic induction
After electricity and magnetism are linked, English scientist Michael Faraday uses electromagnetic induction to generate electricity.

FARADAY'S COIL

Current electricity
Italian inventor Alessandro Volta creates an electric current by stacking discs of zinc, copper, and cardboard soaked in salt water in alternate layers – the first battery.

Timeline of discoveries
Since ancient times, debate and experiment have led to discoveries that further human understanding of how the world works – but there are still many questions left to answer.

VOLTAIC PILE

1700-1890

1847

The Industrial Revolution
Scientific principles understood by the 18th century were applied to large-scale practical machines during the Industrial Revolution. The power of electricity was unlocked, which led the way to a surge in new technology.

NEWCOMEN ENGINE

1712

Steam engine
Thomas Newcomen, an English engineer, builds the world's first practical steam engine. It is followed by James Watt's more efficient engine and Richard Trevithick's steam locomotive.

Static electricity
German scientist Ewald Georg von Kleist invents the Leyden jar, a device that can store a static electric charge and release it later.

LEYDEN JAR

1700-1890

1712

1643   1604

Atmospheric pressure
Italian physicist Evangelista Torricelli creates a simple barometer that demonstrates atmospheric pressure.

BAROMETER

Falling bodies
In a letter to theologian Paolo Sarpi, Italian scientist Galileo outlines his theory that all objects fall at the same rate, regardless of mass or shape.

GALILEO'S EXPERIMENT WITH FALLING BODIES

1600

Earth's magnetism
English scientist William Gilbert theorizes that the Earth must have a huge magnet inside.

1500-1700

Bending light
German monk Theodoric of Freiburg uses bottles of water and water droplets in rainbows to understand refraction.

16TH CENTURY

A new age of science
The scientific revolution, from the mid-16th to the late 18th century, transformed understanding of astronomy and physics. This period saw the development of the scientific method of experiment and observation.

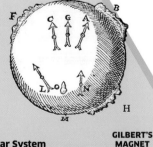

NICOLAUS COPERNICUS

Solar System
Polish astronomer Nicolaus Copernicus states that the Earth and planets orbit round the Sun.

GILBERT'S MAGNET

1300   1543

ENERGY

Energy is all around us – the secret power behind everything in our world, from a bouncing ball to an exploding star. Energy is what makes things happen. It is what gives objects the ability to move, to glow with heat and light, or to make sounds. The ultimate source of all energy on Earth is the Sun. Without energy, there would be no life.

TYPES OF ENERGY

Energy exists in many different forms. They are all closely related and each one can change into other types.

Potential energy
This is stored energy. Climb something, and you store potential energy to jump, roll, or dive back down.

Mechanical energy
Also known as elastic energy, this is the potential stored in stretched objects, such as a taut bow.

Nuclear energy
Atoms are bound together by energy, which they release when they split apart in nuclear reactions.

Chemical energy
Food, fuel, and batteries store energy within the chemical compounds they are made of, which is released by reactions.

Sound energy
When objects vibrate, they make particles in the air vibrate, sending energy waves travelling to our ears, which we hear as sounds.

Heat energy
Hot things have more energy than cold ones, because the particles inside them jiggle around more quickly.

Electrical energy
Electricity is energy carried by charged particles called electrons moving through wires.

Light energy
Light travels at high speed and in straight lines. Like radio waves and X-rays, it is a type of electromagnetic energy.

Kinetic energy
Moving things have kinetic energy. The heavier and faster they are, the more kinetic energy they have.

Measuring energy

Scientists measure energy in joules (J). One joule is the energy transferred to an object by a force of 1 newton (N) over a distance of 1 metre (m), also known as 1 newton metre (Nm).

• Energy of the Sun
The Sun produces four hundred octillion joules of energy each second!

• Energy in candles
A candle emits 80 J – or 80 W – of energy (mainly heat) each second.

• Energy of a light bulb
An LED uses 15 watts (W), or 15 J, of electrical energy each second.

• Energy in water
To raise water temperature 1°C (1.8°F) takes 1 calorie (1/1,000 kilocalories).

• Energy in food
The energy released by food is measured in kilocalories: 1 kcal is 4,184 J.

• Tiny amounts of energy
Ergs measure tiny units of energy. There are 10 million ergs in 1 J.

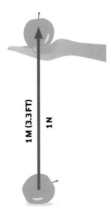
Lifting an apple
One joule is roughly equivalent to lifting an apple 1m (3.3ft).

CONSERVATION OF ENERGY

There's a fixed amount of energy in the Universe that cannot be created or destroyed, but it can be transferred from one object to another and converted into different forms.

Energy conversion

The total amount of energy at the start of a process is always the same at the end, even though it has been converted into different forms. When you switch on a lamp, for example, most of the electrical energy is converted into light energy – but some will be lost as heat energy. However, the total amount of energy that exists always stays the same.

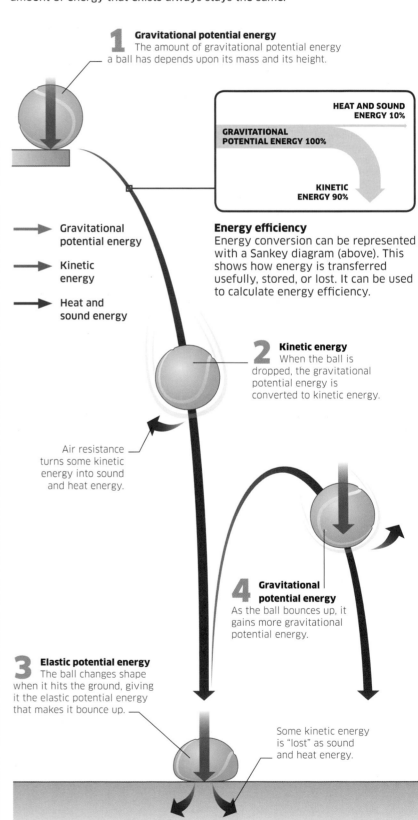

1 Gravitational potential energy
The amount of gravitational potential energy a ball has depends upon its mass and its height.

HEAT AND SOUND ENERGY 10%
GRAVITATIONAL POTENTIAL ENERGY 100%
KINETIC ENERGY 90%

→ Gravitational potential energy

→ Kinetic energy

→ Heat and sound energy

Energy efficiency

Energy conversion can be represented with a Sankey diagram (above). This shows how energy is transferred usefully, stored, or lost. It can be used to calculate energy efficiency.

2 Kinetic energy
When the ball is dropped, the gravitational potential energy is converted to kinetic energy.

Air resistance turns some kinetic energy into sound and heat energy.

4 Gravitational potential energy
As the ball bounces up, it gains more gravitational potential energy.

3 Elastic potential energy
The ball changes shape when it hits the ground, giving it the elastic potential energy that makes it bounce up.

Some kinetic energy is "lost" as sound and heat energy.

ENERGY SOURCES

People in the industrialized world use a lot of energy in homes, business, and industry, for travel and transport. The energy used comes from primary sources such as fossil fuels, nuclear energy, and hydropower. Crude oil, natural gas, and coal are called fossil fuels because they were formed over millions of years by heat from Earth's core and pressure from rock on the remains (fossils) of plants and animals (see p.37).

Energy consumption

Most energy consumed in the USA is from non-renewable sources, with more than 80 per cent derived from fossil fuels. Despite advances, just 10 per cent comes from renewable sources, of which nearly half is from biomass.

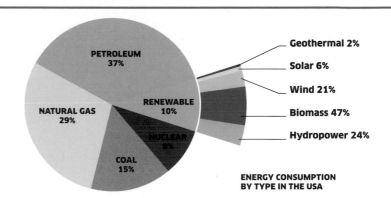

PETROLEUM 37%

NATURAL GAS 29%

RENEWABLE 10%

NUCLEAR 9%

COAL 15%

- Geothermal 2%
- Solar 6%
- Wind 21%
- Biomass 47%
- Hydropower 24%

ENERGY CONSUMPTION BY TYPE IN THE USA

Non-renewable sources

Fossil fuels are limited resources on our planet, which create greenhouse gases (see pp.128–129) and toxic pollutants. Nuclear energy produces fewer greenhouse gases, but leaves harmful waste.

Crude oil
Liquid hydrocarbons found deep underground.

Natural gas
Hydrocarbon gas formed millions of years ago.

Coal
Solid hydrocarbons made by heat and pressure.

Nuclear energy
Energy released by splitting uranium atoms.

Renewable sources

Energy produced by resources that cannot run out, such as sunlight, wind, and water, is more sustainable. Their use does not produce greenhouse gases and other harmful waste products. Biomass releases carbon dioxide, however, and must be offset by planting new trees.

Biomass
Fuel from wood, plant matter, and waste.

Geothermal energy
Heat deep inside the Earth, in water and rock.

Wind power
Moving air caused by uneven heating of Earth.

Solar energy
The Sun's radiation, converted into heat.

Hydropower
The energy of falling or flowing water.

Tidal and wave power
The motion of tides and wind-driven waves.

Energy use

In the developed world, industry and transport are the most energy-hungry sectors, while efficiency has reduced energy consumption in the home.

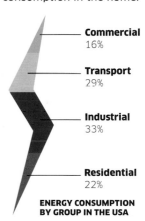

- **Commercial** 16%
- **Transport** 29%
- **Industrial** 33%
- **Residential** 22%

ENERGY CONSUMPTION BY GROUP IN THE USA

ELECTRICAL GRID

Regardless of the primary energy source, most energy is delivered to users as electrical energy. The network of cables used to distribute electricity to homes, offices, and factories is called the electrical grid. Many sources, including wind and solar, feed into the grid, but the majority of electricity is generated in power stations, which use the energy released by burning fossil fuels to power huge electrical generators.

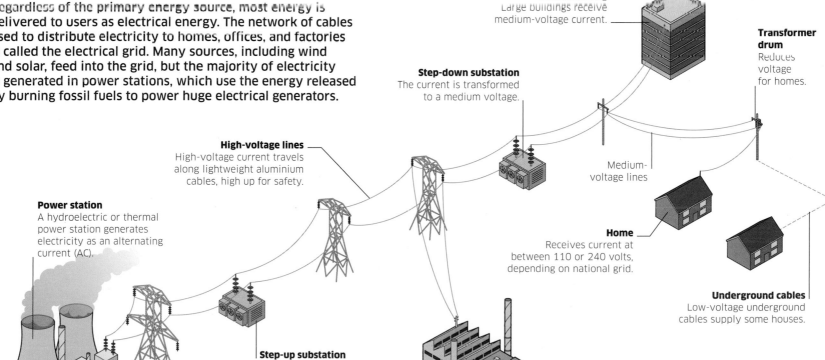

Office building
Large buildings receive medium-voltage current.

Transformer drum
Reduces voltage for homes.

Step-down substation
The current is transformed to a medium voltage.

High-voltage lines
High-voltage current travels along lightweight aluminium cables, high up for safety.

Medium-voltage lines

Power station
A hydroelectric or thermal power station generates electricity as an alternating current (AC).

Home
Receives current at between 110 or 240 volts, depending on national grid.

Underground cables
Low-voltage underground cables supply some houses.

Step-up substation
A transformer boosts the current to a high voltage before it enters the grid.

Factory
Industry receives high-voltage current.

Metals are good **heat conductors** because their electrons are free to move and pass energy on.

Copper, gold, silver, and aluminium are all good **conductors** of heat.

Heat transfer

Heat in this pan of boiling water can be seen to move in three ways – radiation, conduction, and convection – between the heat source, the metal pan, and the water.

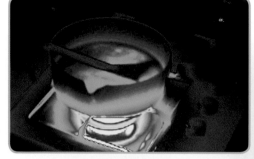

Heat distribution
A thermogram (infrared image) reveals how heat is distributed from the hottest point, the flame, to the coldest, the wooden spoon and stove.

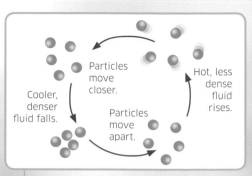

Convection
As a fluid (liquid or gas) heats up, the particles of which it is made move apart, so the fluid becomes less dense and rises. As it moves away from the heat source, the fluid cools down, its density increases, and it falls.

Particles move closer.

Hot, less dense fluid rises.

Cooler, denser fluid falls.

Particles move apart.

Thermal insulation
Materials such as plastic and wood are thermal insulators, which do not conduct heat.

Heat

Heat is energy that increases the temperature of a substance or makes it change state – from a liquid to a gas, for example. Heat can move into or within a substance in three ways: conduction, convection, or radiation.

Atoms and molecules are always moving around. The energy of their movement is called kinetic energy. Some move faster than others, and the temperature of a substance is the average kinetic energy of its atoms and molecules.

Hot particles emit yellow light.

Particles move less away from heat source.

Heat source

Radiation from pan
Some radiant heat is lost from the side of the pan.

Conduction
When the particles (atoms or molecules) of a solid are heated, they move faster, bumping into other particles and making them move faster too. The movement of the particles conducts heat away from the heat source. As the temperature increases in a metal, the particles lose heat as thermal radiation, making the metal glow red, yellow, and then white hot.

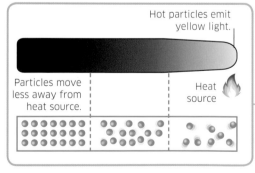

Radiation from flame
Heat moves as radiant energy waves through a gas or vacuum. This is how the Sun heats Earth.

Radiation absorbed by stove
The matt black surface of the stove absorbs some heat radiation.

Heat energy **always** passes from **hot objects** or materials to **cooler ones**.

The **Sun** is the main **source of heat** on Earth.

The **temperature range** on Earth is **less than 150°C (250°F)**.

73

Convection currents in air

In the daytime, warm air rises from the land and cool air flows in from the sea, creating a sea breeze. At night, warm air rises from the sea and cool air flows out to sea.

Land and sea breezes
Currents reverse as the land and sea warm and cool at different rates.

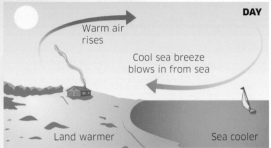

DAY

Warm air rises

Cool sea breeze blows in from sea

Land warmer

Sea cooler

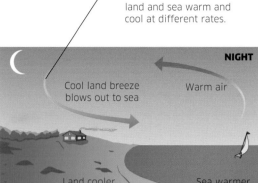

NIGHT

Cool land breeze blows out to sea

Warm air

Land cooler

Sea warmer

Measuring temperature

Temperature measures how hot or cold an object is by taking the average value of its heat energy. It is measured in degrees Celsius (°C), Fahrenheit (°F), or Kelvin (K). A degree is the same size on the °C scale and K scale. All atoms stop moving at absolute zero (0K).

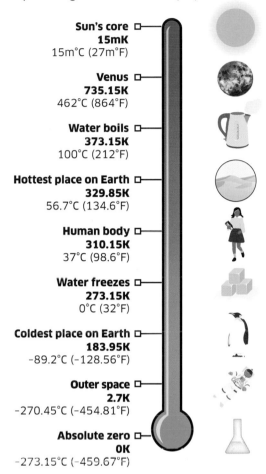

Sun's core
15mK
15m°C (27m°F)

Venus
735.15K
462°C (864°F)

Water boils
373.15K
100°C (212°F)

Hottest place on Earth
329.85K
56.7°C (134.6°F)

Human body
310.15K
37°C (98.6°F)

Water freezes
273.15K
0°C (32°F)

Coldest place on Earth
183.95K
-89.2°C (-128.56°F)

Outer space
2.7K
-270.45°C (-454.81°F)

Absolute zero
0K
-273.15°C (-459.67°F)

Steel
The steel lining the pan is a less good heat conductor than copper, but also less reactive, so it is slower to corrode.

Copper
The copper exterior of the pan is a good conductor of heat, but corrodes easily.

Radiation reflected off pan
The shiny metal exterior absorbs heat radiation from the flame, but also reflects some back.

Heat loss and insulation

Heat is easily lost from our homes through floors, walls, and roofs, windows and doors. To increase energy efficiency by reducing heat loss, materials that are poor conductors, such as plastics, wood, cork, fibreglass, and air, can be used to provide insulation.

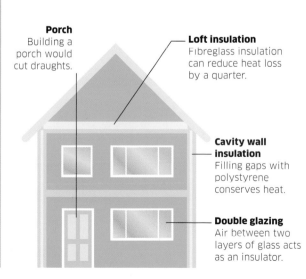

Porch
Building a porch would cut draughts.

Loft insulation
Fibreglass insulation can reduce heat loss by a quarter.

Cavity wall insulation
Filling gaps with polystyrene conserves heat.

Double glazing
Air between two layers of glass acts as an insulator.

74 energy and forces ∘ **NUCLEAR ENERGY**

449 The number of **operational nuclear reactors in the world**, with many more being built.

Nuclear energy

Nuclear reactions are a highly efficient way of releasing energy. Smashing atomic particles together sets off a chain reaction – producing enough heat to generate large amounts of electricity.

Most elements have several slightly different forms, called isotopes. Each isotope of an element has a different number of neutrons. Radioactive isotopes have too many or too few neutrons, making them unstable. Isotopes of heavy elements, such as uranium and plutonium, may break apart, or decay, producing radiation. Atomic nuclei can also be broken apart (fission) or joined together (fusion) artificially to release energy, which can be harnessed in nuclear power stations and weapons.

Nuclear reactor

Nuclear fission power stations are found all over the world. They all use the same basic principles to generate electricity. Firstly, atoms are smashed apart in the reactor to release heat energy. This energy passes into a nearby chamber to heat up water and produce large quantities of steam. The steam powers spinning turbines attached to a generator, which converts this kinetic energy into the electricity that is pumped out to the world.

Turbines
A series of turbines are spun round by the steam.

Steam
The water heated in the tank evaporates into steam, which passes along pipes to the turbines.

Protective dome
A concrete dome around the reactor absorbs radiation.

Control rods
Control rods lowered into the core slow the reaction by absorbing excess neutrons.

Fuel rods
Rods of nuclear fuel are lowered to start a fission reaction.

Reactor core
Atomic nuclei split inside the reactor, releasing heat energy.

Heated water
Water inside the reactor is heated as the reaction takes place.

Inner loop
Water from the reactor heats up a tank of water, before flowing back into the reactor.

Outer loop
Water from the turbine unit returns to the steam generator, ready to be heated again.

Cherenkov radiation
The atomic particles in the reactor travel incredibly fast. In doing so, they generate a type of radiation called Cherenkov radiation, which makes the water surrounding the reactor glow an eerie blue colour.

Types of radiation

When unstable nuclei break apart, or decay, they may release three types of radiation: alpha, beta, and gamma. Alpha and beta radiation are streams of particles released by atomic nuclei. Gamma rays, released during alpha and beta decay or even by lightning, are a form of electromagnetic radiation – similar to light, but more powerful and dangerous.

British physicists **John Cockcroft and Ernest Walton** carried out the **first artificial nuclear fission** in 1932.

11 per cent of the world's electricity is provided by nuclear power plants.

75

Electricity pylons
These carry power lines that transmit electricity from the power station to electricity users.

Generator
The generator converts energy from the turbines into electricity.

Condenser loop
Cooled water is pumped back to the turbine, ready to be heated again.

Cooling towers
Large towers receive the steam and condense it back into water.

Nuclear fission

The nuclei of atoms can split apart or join together, forming new elements and releasing energy. A large atomic nucleus splitting in two is called nuclear fission. A neutron hits the nucleus of a uranium atom, causing it to split, or fission, in two. More neutrons are released as a result, and these hit more nuclei, creating a chain reaction. The extra energy that is released ends up as heat that can be used to generate electricity.

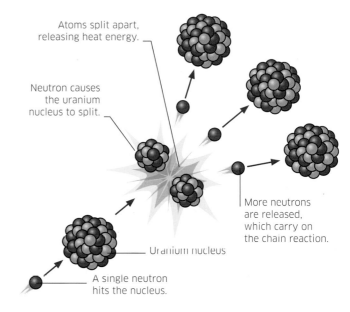

Atoms split apart, releasing heat energy.

Neutron causes the uranium nucleus to split.

More neutrons are released, which carry on the chain reaction.

Uranium nucleus

A single neutron hits the nucleus.

Nuclear fusion

The process in which two smaller atomic nuclei join together is called fusion. Two isotopes of hydrogen are smashed into each other to make helium, releasing heat energy and a spare neutron. Fusion takes place in stars, but has not yet been mastered as a viable form of producing energy on Earth, due to the immense heat and pressure needed to start the process.

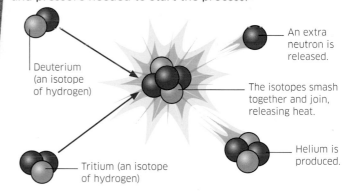

Deuterium (an isotope of hydrogen)

An extra neutron is released.

The isotopes smash together and join, releasing heat.

Tritium (an isotope of hydrogen)

Helium is produced.

Alpha radiation
Some large nuclei release a positively charged particle made of two protons and two neutrons, called an alpha particle.

Beta radiation
In some nuclei, a neutron changes to a proton, creating an electron called a beta particle, which shoots out of the nucleus.

Gamma radiation
Gamma rays are electromagnetic waves released during alpha and beta decay.

Containing radiation

Radiation can be extremely harmful to human health and containing it can be tricky. Alpha, beta, and gamma radiation can pass through different amounts of matter because they have different speeds and energy. Alpha particles can be stopped by skin, or just a sheet of paper. Beta rays can pass through skin but not metal. Gamma rays can only be stopped by a sheet of lead or thick concrete.

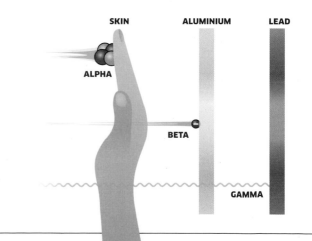

SKIN   ALUMINIUM   LEAD

ALPHA

BETA

GAMMA

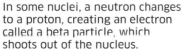

Acoustic guitar sounds

When a player plucks the strings of a guitar, each string vibrates at a different frequency to produce a note of a different pitch – higher- or lower-sounding. The pitch of the note produced depends on the length, tension, thickness, and density of the string. The strings' vibration passes into the body of the instrument, which causes air inside and outside it to vibrate, making a much louder sound.

String thickness and density
The thickness of strings affects their frequency and pitch: the thickest string creates the lowest frequency and lowest-pitch notes. Strings made of more dense materials will have a lower pitch.

Sound waves
Vibrating air comes out of the sound hole in waves that spread out evenly in all directions like the ripples in a pool of water.

Sound hole
Air at the sound hole oscillates, adding resonance.

Soundboard
The large surface area of the soundboard (also known as the top plate) vibrates, creating sound energy.

Saddle and bridge
The string vibrations are transmitted to the saddle and bridge of the guitar.

Vibrating soundboard

Vibrating air

Braces
On the inside of the top and back plates, braces add structural support to the guitar. The geometric pattern they are arranged in affects the sound made.

Hollow body
The hollow body amplifies sound energy travelling through the guitar.

Sounds louder than **85 decibels** can **damage human hearing**.

Ultrasonic waves have a **frequency** higher than audible sound waves.

77

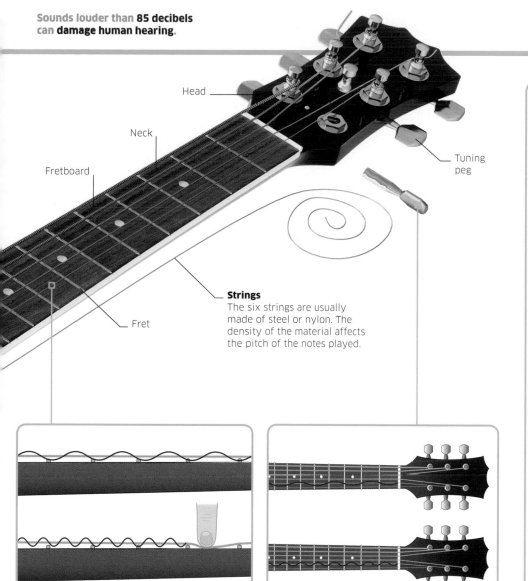

Head

Neck

Fretboard

Tuning peg

Fret

Strings
The six strings are usually made of steel or nylon. The density of the material affects the pitch of the notes played.

String length
Frets (raised bars) are spaced along the fretboard on the front of the neck. The player presses a string down on the fretboard to shorten its length, increasing the frequency and raising the pitch of the sound.

String tension
Turning the tuning pegs enables the player to tighten or loosen the strings, adjusting the pitch so that the guitar is in tune. As the strings are tightened, the frequency increases, raising the pitch.

Sound

Sound carries music, words, and other noises at high speed. It travels in waves, created by the vibration of particles within a solid, liquid, or gas.

If you pluck a guitar string, it vibrates. This disturbs the air around it, creating a wave of high and low pressure that spreads out. When the wave hits our ears, the vibrations are passed on to tiny hairs in the inner ear, which send information to the brain, where it is interpreted. What distinguishes sounds such as human voices from one another is complex wave shapes that create distinctive timbre and tone.

20 Hz to 20 kHz – the normal range of human hearing. This range decreases as people get older. Children can usually hear higher frequencies than adults.

How sound travels
Sounds waves squeeze and stretch the air as they travel. They are called longitudinal waves because the particles of the medium they are travelling through vibrate in the direction of the wave.

Vibrating particles
As vibrations travel through air, particles jostle each other to create high-pressure areas of compression and areas of low-pressure rarefaction.

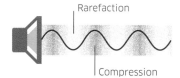

Rarefaction

Compression

Loudness and amplitude
The energy of a sound wave is described by its amplitude (height from centre to crest or trough), corresponding to loudness.

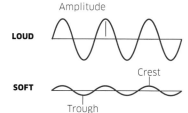

Amplitude

LOUD

Crest

SOFT

Trough

Pitch and frequency
A sound wave's pitch is defined by its frequency: the number of waves that pass a point in a given time. It is measured in hertz (Hz).

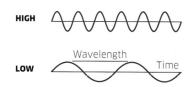

HIGH

Wavelength

Time

LOW

Speed of sound in different materials
Sound moves fastest in solids, because the particles are closer together, and slowest in gases, such as air, because the particles are further apart. The speed of sound is measured in metres per second.

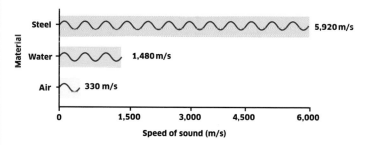

Steel — 5,920 m/s
Water — 1,480 m/s
Air — 330 m/s

Material

0   1,500   3,000   4,500   6,000

Speed of sound (m/s)

The decibel range
Loudness describes the intensity of sound energy, and is measured in decibels (dB) on a logarithmic scale, so 20 dB is 10 times more intense than 10 dB, or twice as loud. Human hearing ranges from 0 to 150 dB.

LEAF FALLING NEARBY (10 dB)
Barely audible

WHISPERING IN EAR (30 dB)
Quiet

SPEAKING NEAR YOU (60 dB)
Moderate

VIOLIN AT ARM'S LENGTH (90 dB)
Loud

FRONT OF ROCK GIG (120 dB)
Very loud

FIREWORK AT CLOSE RANGE (150 dB)
Painfully loud

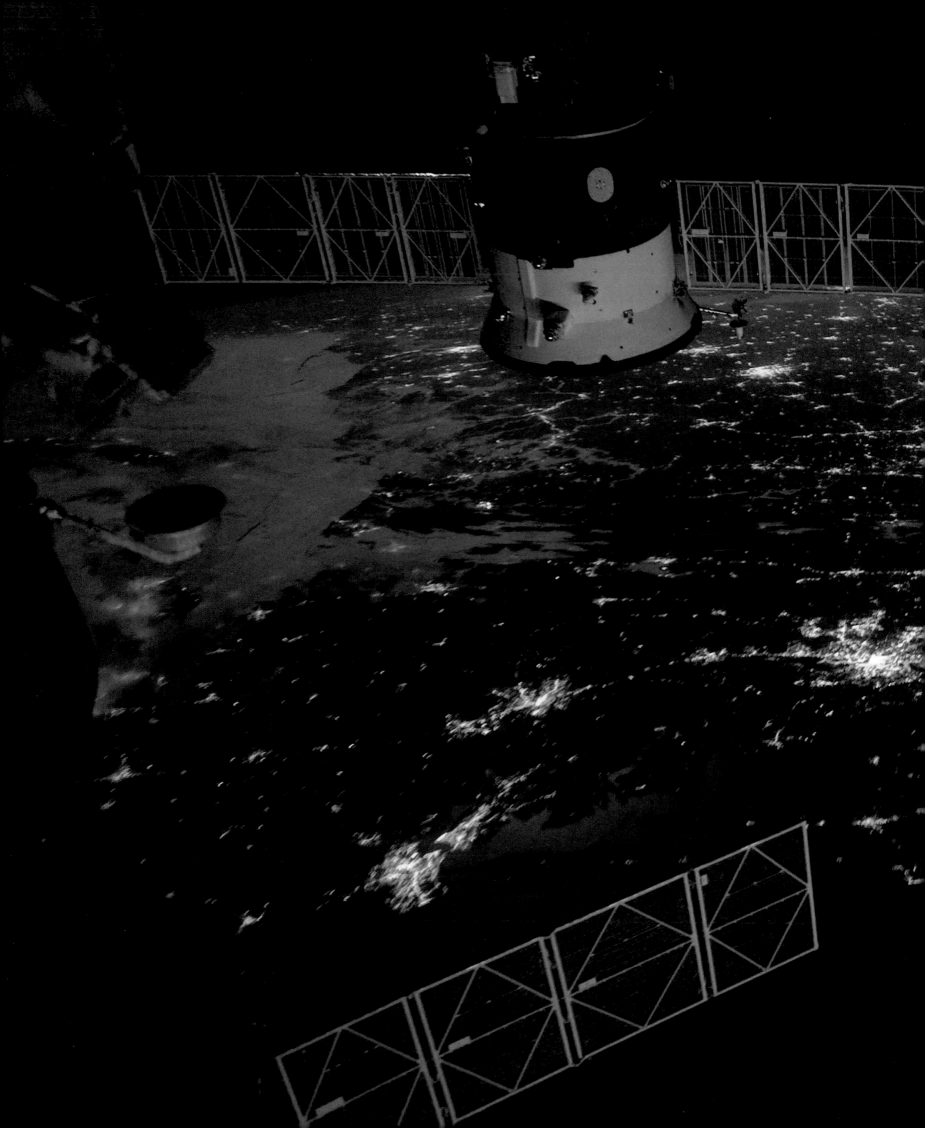

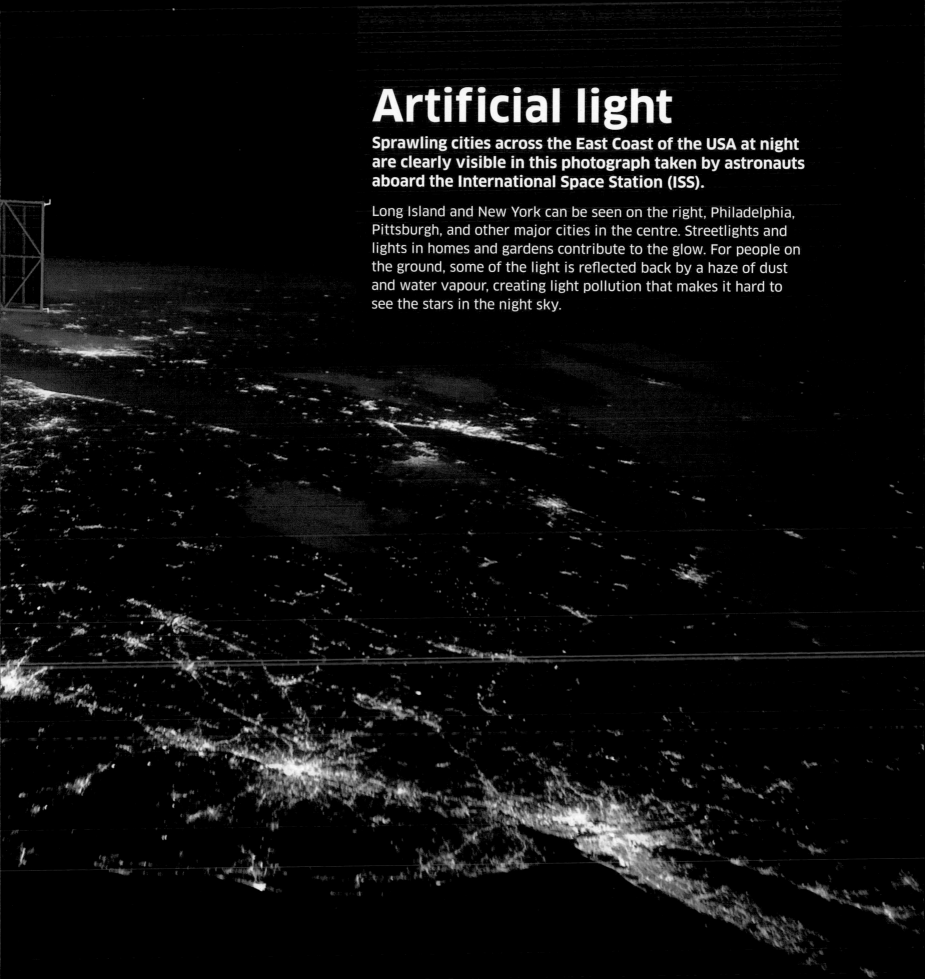

Artificial light

Sprawling cities across the East Coast of the USA at night are clearly visible in this photograph taken by astronauts aboard the International Space Station (ISS).

Long Island and New York can be seen on the right, Philadelphia, Pittsburgh, and other major cities in the centre. Streetlights and lights in homes and gardens contribute to the glow. For people on the ground, some of the light is reflected back by a haze of dust and water vapour, creating light pollution that makes it hard to see the stars in the night sky.

Gamma rays
The highest-energy waves, with wavelengths the size of an atomic nucleus, gamma rays are emitted by nuclear fission in weapons and reactors and by radioactive substances. Gamma radiation is very harmful to human health.

Visible light
This is the range of wavelengths that is visible to the human eye. Each drop in a raindrop is like a tiny prism that splits white light into the colours of the spectrum.

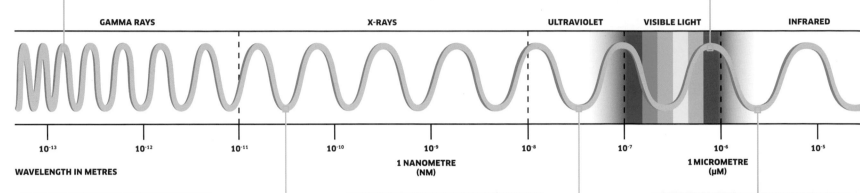

| GAMMA RAYS | X-RAYS | ULTRAVIOLET | VISIBLE LIGHT | INFRARED |

10^{-13} 10^{-12} 10^{-11} 10^{-10} 10^{-9} 10^{-8} 10^{-7} 10^{-6} 10^{-5}

WAVELENGTH IN METRES

1 NANOMETRE
(NM)

1 MICROMETRE
(µM)

X-rays
With the ability to travel through soft materials but not hard, dense ones, X-rays are used to look inside the body and for security bag checks.

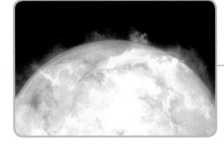

Ultraviolet (UV)
Found in sunlight, UV radiation can cause sunburn and eye damage. The shortest, most harmful wavelengths are blocked by the ozone layer.

Infrared
Known as heat radiation, infrared is invisible, but special cameras are able to detect it and "see" the temperature of objects such as these penguins.

Electromagnetic radiation

Light is one of several types of wave energy called electromagnetic radiation, which also includes radio waves, X-rays, and gamma radiation.

Electromagnetic radiation reaches us from the Sun, stars, and distant galaxies. The Earth's atmosphere blocks most types of radiation, but allows radio waves and light, which includes some wavelengths of infrared and ultraviolet, to pass through.

The electromagnetic spectrum beyond visible light was discovered between 1800, when British astronomer William Herschel first observed infrared, and 1900, when French physicist Paul Villard discovered gamma radiation.

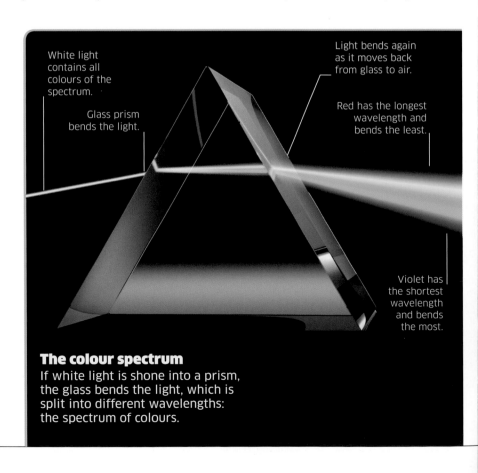

White light contains all colours of the spectrum.

Glass prism bends the light.

Light bends again as it moves back from glass to air.

Red has the longest wavelength and bends the least.

Violet has the shortest wavelength and bends the most.

The colour spectrum
If white light is shone into a prism, the glass bends the light, which is split into different wavelengths: the spectrum of colours.

All electromagnetic waves travel through space at the **speed of light**, which is **299,792,458 m/s** (commonly rounded to 300,000 km/s).

Extremely low frequency (ELF) radio waves are used to **communicate with submarines.**

81

Microwaves
On Earth, microwaves are used for radar, cell-phone, and satellite communications. Scientists have captured images (left) of microwaves left over from the Big Bang at the birth of the Universe.

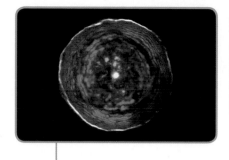

Radio waves
The longest waves on the spectrum, radio waves carry TV as well as radio signals. Radio telescopes are able to capture radio waves emitted by sources in space and convert them into images, such as this star (left).

Waves at this end of the spectrum have least energy and lowest frequencies.

MICROWAVE

RADIO WAVES

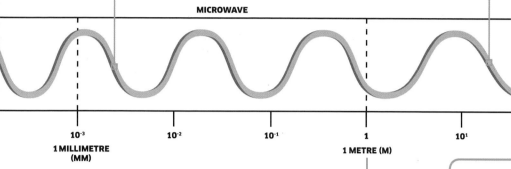

| 10^{-3} | 10^{-2} | 10^{-1} | 1 | 10^{1} | 10^{2} | 10^{3} | 10^{4} | 10^{5} |

1 MILLIMETRE (MM)

1 METRE (M)

1 KILOMETRE (KM)

The electromagnetic spectrum

There are electromagnetic waves over a wide range of wavelengths, from gamma waves, which have the shortest wavelength and highest energy, to radio waves, which have the longest wavelength and lowest energy. All electromagnetic waves are invisible, except for those that make up light. As you move along the spectrum, different wavelengths are used for a variety of tasks, from sterilizing food and medical equipment to communications.

The dividing line between some types of electromagnetic radiation is distinct, whereas other types overlap. Microwaves, for example, are the shortest wavelength radio waves, ranging from 1 mm to 1 m.

Electromagnetic waves

All types of electromagnetic radiation are transverse waves that transfer energy from place to place and can be emitted and absorbed by matter. Electromagnetic radiation travels as waves of electric and magnetic fields that oscillate (vibrate) at right angles to each other and to the direction of travel.

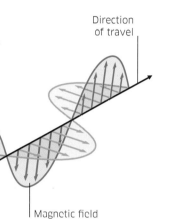

Direction of travel

Electric field

Magnetic field

Seeing colour

We see colour based on information sent to the brain from light-sensitive cells in the eye called cones. There are three types of cone, which respond to red, green, or blue light. We see all colours as a mix of these three colours. Objects reflect or absorb the different colours in white light. We see the reflected colours.

White object
White objects reflect all the colours that make up the visible light spectrum, which is why they appear white.

Black object
Black objects absorb all the colours of the visible light spectrum and reflect none. They also absorb more heat.

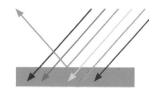

Green object
We see green objects because they reflect only the green wavelengths of visible light.

Light scattering

When sunlight hits Earth's atmosphere, air molecules, water droplets, and dust particles scatter the light, but they don't scatter the colours equally. This is why the sky is blue, clouds are white, and sunsets red.

Blue sky
The blue of the sky is caused by air molecules in the atmosphere, which scatter short-wavelength light at the blue end of the spectrum. Larger water and dust particles scatter the full spectrum as white light. The bluer the sky, the purer the air.

Red sunset
When the Sun is low in the sky, light takes a longer path through the atmosphere, more light is scattered, and shorter wavelengths are absorbed. At sunrise and sunset, clouds may appear red or orange, reflecting the colour of light shining on them.

Telephone network

Mobile phones connect to base stations, each providing coverage of a hexagonal area called a cell. Each cell has a number of frequencies or channels available to callers. As mobiles each connect to a particular base station, the same frequencies can be used for callers using base stations elsewhere. Landline calls go through local and main exchanges.

Calling from a moving mobile
User A's call is given a channel and routed via a base station to the mobile exchange. User A's phone checks the signal strength from nearby base stations, feeding this information back to the mobile exchange. It indicates the current signal is weakening as the caller leaves the cell.

Call handed over to new cell
The mobile exchange readies a new channel for user A in the cell they are moving to and sends this information to user A's phone. User A's phone signals to the new base station its arrival in the new cell and the old channel is shut down.

Moving mobile call received
The mobile exchange scans for user B and puts through the call. B should not notice when A's signal is handed over.

Satellite phone
Instead of linking to base towers, these phones send a high-frequency signal to the nearest satellite, which bounces it back to a main exchange.

International exchange
Calls to other countries are routed through the caller's main exchange and on to an international exchange.

Relay tower
Radio links at microwave frequencies connect more distant exchanges via high relay towers.

1 Caller dials landline number
The mobile connects by microwave to a nearby base station.

2 Base station in cell
The base station routes the call to a mobile exchange. Each cell has a base station that sends and receives signals at a range of frequencies. Dense urban areas have more, smaller cells to cope with user demand.

3 Mobile exchange
The mobile exchange passes the call to the main exchange. Mobile exchanges receive signals from many base stations.

4 Main exchange
The main exchange transfers the call to the local exchange. Local exchanges across a wide area are all connected to a main exchange.

5 Local exchange
The call is routed from the local exchange to a landline. All the telephones in a small area are connected to the local exchange.

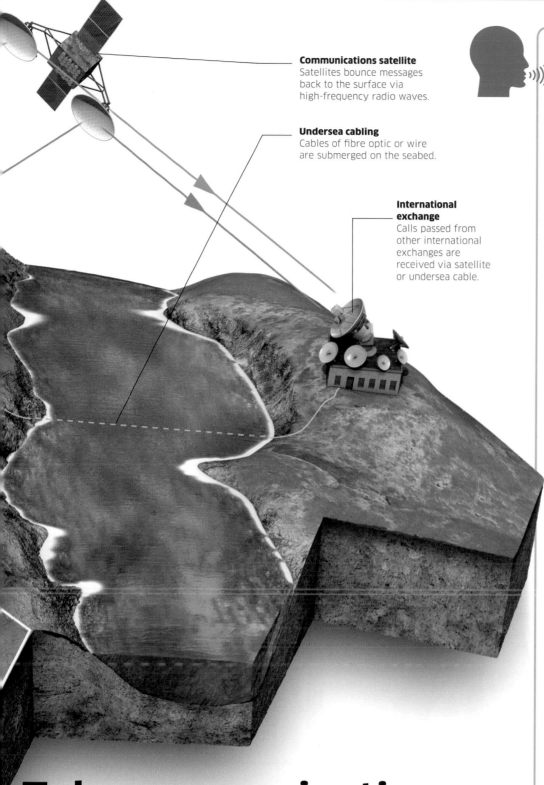

Communications satellite
Satellites bounce messages back to the surface via high-frequency radio waves.

Undersea cabling
Cables of fibre optic or wire are submerged on the seabed.

International exchange
Calls passed from other international exchanges are received via satellite or undersea cable.

Speech conversion
Our voices are converted from analogue signals to digital ones to make calls.

1 A mobile phone captures sound as a continuously varying or analogue signal. The signal is measured at various points and each point is given a value. Here a point on the signal is measured as 3, which is shown as its binary equivalent of 0011. The phone's analogue to digital converter produces strings of these binary numbers (see p.95).

2 The 1s and 0s of the binary number 0011 become off/off/on/on. The phone transmits the on/off values, encoding them as sudden changes to the signal's waves. The signal passes from base station to mobile exchange to base station.

3 The phone receives the digital signal and interprets the on/off transmission as strings of binary numbers. The phone's digital to analogue converter turns the binary numbers back into analogue information.

4 The phone's speaker sends an analogue signal we hear as a sound wave.

OFF OFF ON ON
0 0 1 1

OFF OFF ON ON

OFF OFF ON ON
0 0 1 1

Telecommunications

Modern telecommunications use electricity, light, and radio as signal carriers. The global telephone network enables us to communicate worldwide, using radio links, fibre-optic cables, and metal cables.

Signals representing sounds, images, and other data are sent as either analogue signals, which are unbroken waves, or as digital signals that send binary code as abrupt changes in the waves. Radio waves transmit radio and TV signals through the air around Earth, while microwave wavelengths are used in mobile phones, Wi-Fi, and Bluetooth. Cables carry signals both above and below ground – as electric currents along metal wires, or as pulses of light that reflect off the glass interiors of fibre-optic cables.

The ionosphere and radio waves
The ionosphere is a region of the atmosphere that contains ions and free electrons. This causes it to reflect some lower-frequency, longer-wavelength radio waves over large distances.

Short and medium waves are reflected off the top of the ionosphere and the Earth's surface.

Communications satellites retransmit signals to more distant areas.

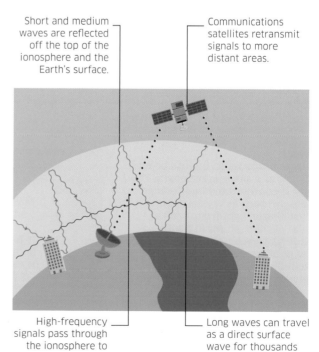

High-frequency signals pass through the ionosphere to reach satellites.

Long waves can travel as a direct surface wave for thousands of kilometres.

Types of telescope

Refracting telescopes use lenses to gather and focus light. Reflecting telescopes do the same with mirrors – huge space telescopes use very large mirrors. Compound telescopes combine the best of lenses and mirrors.

Refracting telescope

A large convex lens focuses light rays to a mirror that reflects the light into the eyepiece, where a lens magnifies the image. Lenses refract the light, causing colour distortion.

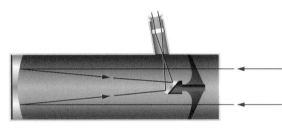

Reflecting telescope

A concave mirror reflects and focuses light to a secondary mirror, which reflects it into an eyepiece, where a lens magnifies the image. There is no colour distortion.

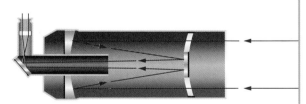

Compound telescope

The most common type of telescope, this combines lenses and mirrors to maximize magnification and eliminate distortion.

Telescopes

Powerful telescopes make faint objects, such as distant stars and galaxies, easier to see. They work by first gathering as much light as they can, using either a lens or a mirror, and then focusing that light into a clear image.

There are two main types of telescope: refracting, which focus light using lenses, and reflecting, which focus light using mirrors. Optical telescopes see visible light, but telescopes can also look for different kinds of electromagnetic radiation: radio telescopes receive radio waves and X-ray telescopes image X-ray sources. Telescopes use large lenses compared to microscopes, which are used to look at things incredibly close up, while binoculars work like two mini telescopes side by side.

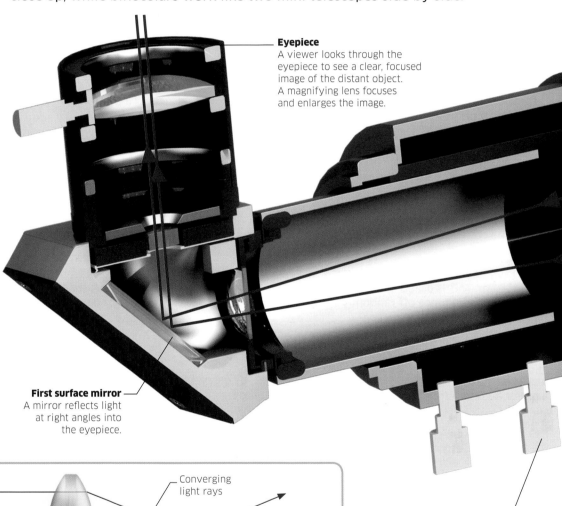

Eyepiece
A viewer looks through the eyepiece to see a clear, focused image of the distant object. A magnifying lens focuses and enlarges the image.

First surface mirror
A mirror reflects light at right angles into the eyepiece.

Focus knob
Twisting the knob adjusts the focal length to focus the image.

Convex and concave lenses

Convex, or converging, lenses take light and focus it into a point behind the lens, called the principal focus. This is the type of lens used in the glasses of a short-sighted person. By contrast, concave, or diverging, lenses spread light out. When parallel rays pass through a concave lens, they diverge as if they came from a focal point, known as the principal focus, in front of the lens.

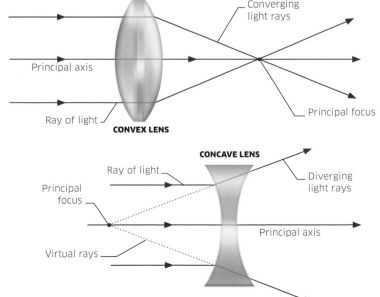

Converging light rays

Principal axis

Ray of light

Principal focus

CONVEX LENS

CONCAVE LENS

Ray of light

Principal focus

Diverging light rays

Principal axis

Virtual rays

A GERMAN-DUTCH LENS MAKER **CALLED HANS LIPPERSHEY DEVELOPED THE EARLIEST** REFRACTING TELESCOPE **IN THE YEAR 1608.** GALILEO IMPROVED THE DESIGN.

Isaac Newton created the first reflecting telescopes to get around the problem of colour distortion.

The **Hubble Space Telescope** can see as far as **13 billion light years away**, to a distant galaxy called **MACS0647-JD**.

When you look at a star **5,000 light years away**, you are looking **back in time** as the light you are seeing left the star **5,000 years ago**.

87

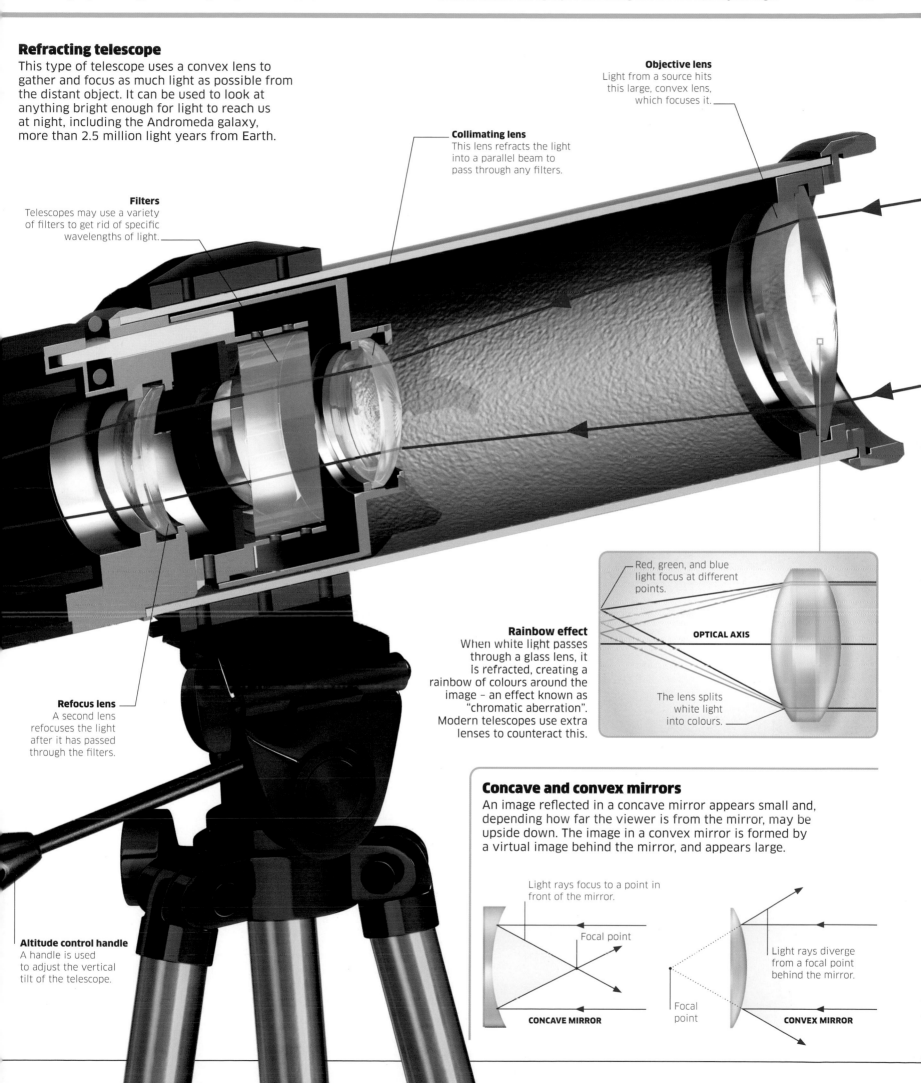

Refracting telescope

This type of telescope uses a convex lens to gather and focus as much light as possible from the distant object. It can be used to look at anything bright enough for light to reach us at night, including the Andromeda galaxy, more than 2.5 million light years from Earth.

Objective lens
Light from a source hits this large, convex lens, which focuses it.

Collimating lens
This lens refracts the light into a parallel beam to pass through any filters.

Filters
Telescopes may use a variety of filters to get rid of specific wavelengths of light.

Refocus lens
A second lens refocuses the light after it has passed through the filters.

Altitude control handle
A handle is used to adjust the vertical tilt of the telescope.

Rainbow effect
When white light passes through a glass lens, it is refracted, creating a rainbow of colours around the image – an effect known as "chromatic aberration". Modern telescopes use extra lenses to counteract this.

Red, green, and blue light focus at different points.

OPTICAL AXIS

The lens splits white light into colours.

Concave and convex mirrors

An image reflected in a concave mirror appears small and, depending how far the viewer is from the mirror, may be upside down. The image in a convex mirror is formed by a virtual image behind the mirror, and appears large.

Light rays focus to a point in front of the mirror.

Focal point

CONCAVE MIRROR

Light rays diverge from a focal point behind the mirror.

Focal point

CONVEX MIRROR

Unlike poles attract, like poles repel
The invisible field of force around a magnet is called a magnetic field. Iron filings show how the magnetic field loops around the magnet from pole to pole.

Attraction
Unlike or opposite poles (a north pole and a south pole) attract each other. Iron filings reveal the lines of force running between unlike poles.

Repulsion
Like poles (two north or two south poles) repel each other. Iron filings show the lines of force being repelled between like poles.

Magnetic induction
An object made of a magnetic material, such as a steel paper clip, is made of regions called domains, each with its own magnetic field. A nearby magnet will align the domain's fields, turning the object into a magnet. The two magnets now attract each other - that is why paper clips stick to magnets. Stroking a paper clip with a magnet can align the domains permanently.

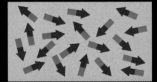

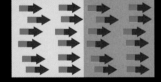

Domains scattered
In an unmagnetized object, the domains point in all directions.

Domains aligned
When a magnet is nearby, the domain's fields align in the object.

Magnetism
Magnetism is an invisible force exerted by magnets and electric currents. Magnets attract iron and a few other metals, and attract or repel other magnets. Every magnet has two ends, called its north and south poles, where the forces it exerts are strongest.

A magnetic material can be magnetized or will be attracted to a magnet. Iron, cobalt, nickel and their alloys, and rare earth metals are all magnetic, which means they can be magnetized by stroking with another magnet or by an electric current. Once magnetized, these materials stay magnetic unless demagnetized by a shock, heat, or an electromagnetic field (see p.93). Most other materials, including aluminium, copper, and plastic, are not magnetic.

Magnetic compass
Made of magnetized metal and mounted so that it can spin freely, the needle of a magnetic compass lines up in a north–south direction in Earth's magnetic field. Because the Earth's magnetic North Pole attracts the north, or north-seeking, pole of other magnets, it is in reality the south pole of our planet's magnetic field.

A teardrop-shaped magnetic field
Earth's magnetic field protects us from the harmful effects of solar radiation. In turn, a stream of electrically charged particles from the Sun, known as the solar wind, distorts the magnetic field into a teardrop shape and causes the auroras - displays of light around the poles (see pp.90–91).

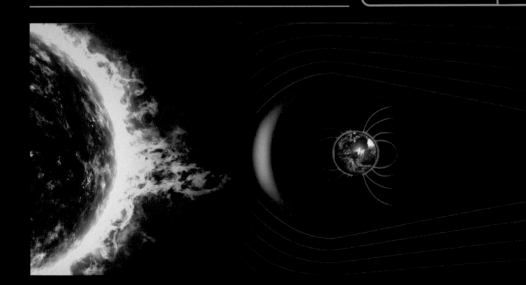

Distortion of the magnetosphere
The stream of charged particles from the Sun compresses Earth's magnetic field on the side nearest the Sun and draws the field away from Earth into a long "magnetotail" on the far side.

Magnetic fields by themselves are **invisible to the human eye**.

The strongest magnets are **rare earth** magnets, made from **neodymium**.

Earth's inner core is believed to be an **alloy of magnetic iron and nickel**.

89

Magnetic and geographic north

There is a difference of a few degrees between the direction that a compass points, known as true north, and the geographic North Pole, which is on the axis of rotation that Earth spins around as it orbits the Sun. In reality, the magnetic poles are constantly moving, and reverse completely every few thousand years.

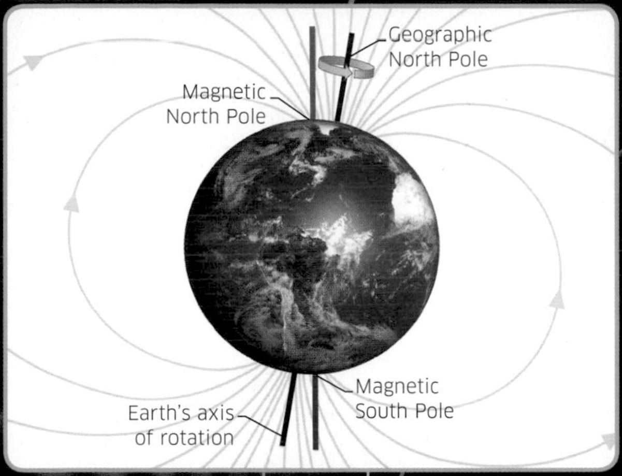

Geographic North Pole

Magnetic North Pole

Magnetic South Pole

Earth's axis of rotation

Earth's magnetism

The Earth can be thought of as one big, powerful magnet with a magnetic force field, called the magnetosphere, that stretches thousands of kilometres into space. The magnetic field is produced by powerful electric currents in the liquid iron and nickel swirling around in Earth's outer core.

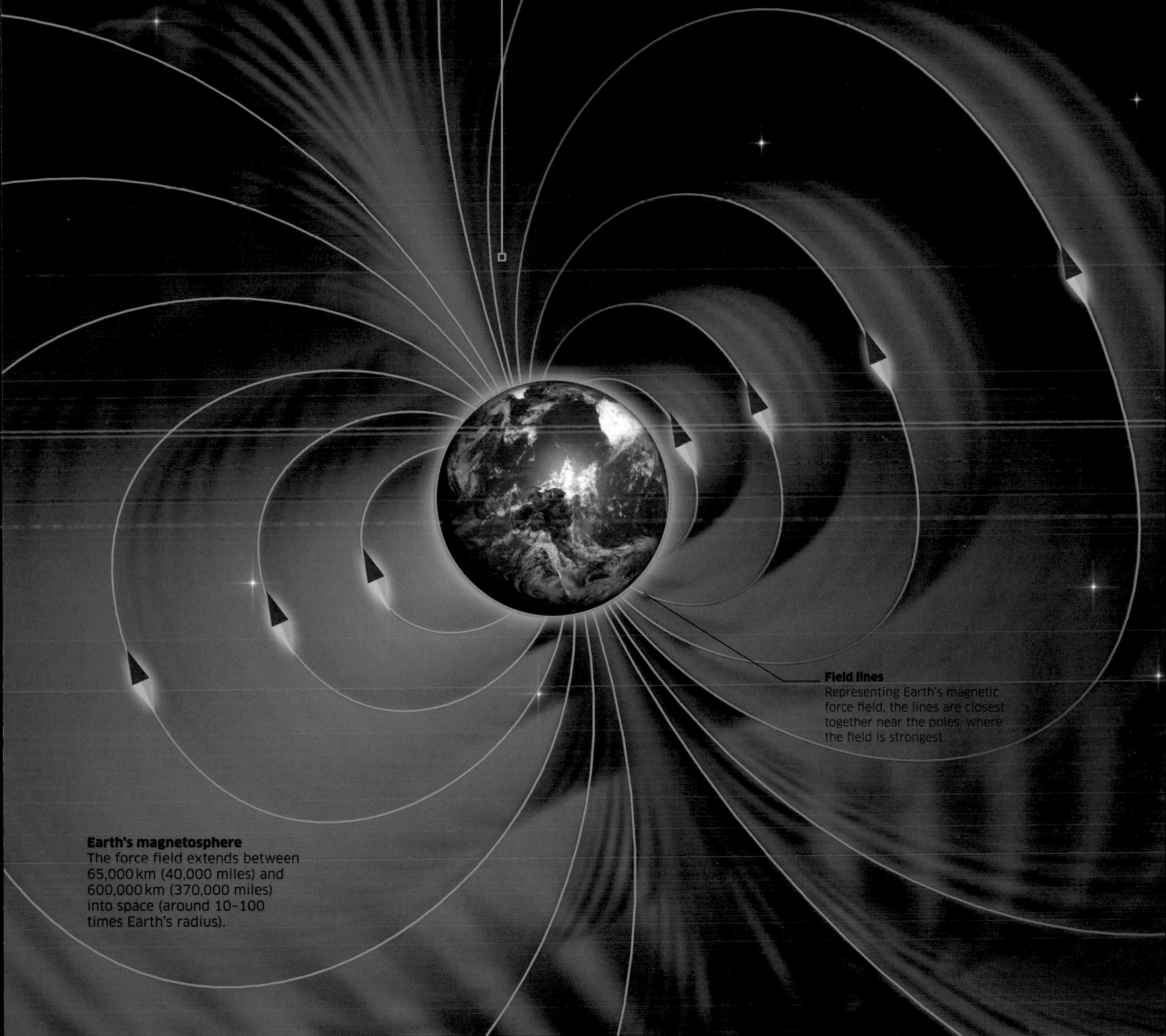

Field lines

Representing Earth's magnetic force field, the lines are closest together near the poles, where the field is strongest.

Earth's magnetosphere

The force field extends between 65,000 km (40,000 miles) and 600,000 km (370,000 miles) into space (around 10–100 times Earth's radius).

Aurora borealis

The spectacular natural light show known as the aurora borealis, or northern lights, is a dazzling spectacle of ribbons and sheets of green, yellow, and pink light.

The cause of the aurora is a stream of charged particles ejected from the surface of the Sun, known as the solar wind. These particles are guided towards the poles by Earth's magnetic field. When they hit oxygen and nitrogen molecules in the atmosphere, electrons in the molecules emit coloured light. The northern lights - and aurora australis, or southern lights, around the South Pole - occur whenever the solar wind blows - typically about two hundred nights a year.

94 energy and forces ∘ **ELECTRONICS**

In 1965, Intel co-founder George Moore correctly predicted that the number of transistors on a chip would double every two years.

Electronics

Electric current is caused by a drift of electrons through a circuit. An electronic device uses electricity in a more precise way than simple electric appliances, to capture digital photos or play your favourite songs.

While it takes a large electric current to boil a kettle, electronics uses carefully controlled electric currents thousands or millions of times smaller, and sometimes just single electrons, to operate a range of complex devices. Computers, smartphones, amplifiers, and TV remote controls all use electronics to process information, communicate, boost sound, or switch things on and off.

Printed circuit board (PCB)
The "brain" of a smartphone is on its printed circuit board – a pre-manufactured electronic circuit unique to a particular device. The PCB is made from interconnected chips (left), each of which is constructed from a tiny wafer of silicon and has an integrated circuit inside it containing millions of microscopic components.

Motherboard
The main printed circuit board, which is the phone's main processor, is also referred to as a mainboard or logic board.

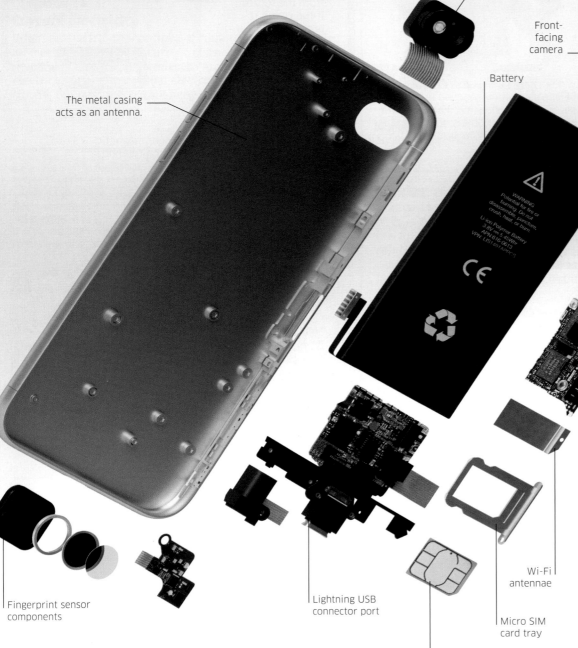

Digital camera

Front-facing camera

Battery

The metal casing acts as an antenna.

Fingerprint sensor components

Lightning USB connector port

Micro SIM card

Micro SIM card tray

Wi-Fi antennae

Electronic components
Electronic circuits are made of building blocks called components. A transistor radio may have a few dozen, while a processor and memory chip in a computer could have billions. Four components are particularly important and appear in nearly every circuit.

Diode
Diodes make electric current flow in just one direction, often converting alternating to direct current.

Resistor
Resistors reduce electric current so it is less powerful. Some are fixed and others are variable.

Transistor
Transistors switch current on and off or convert small currents into bigger ones.

Capacitor
Capacitors store electricity. They are used to detect key presses on touch screens.

Smartphone
Mobile phones are now so advanced that they are really handsized computers. As well as linking to other digital devices, they contain powerful processor chips and plenty of memory to store applications.

Smartphones today are more powerful than the NASA computers that sent Apollo 11 to the Moon.

95

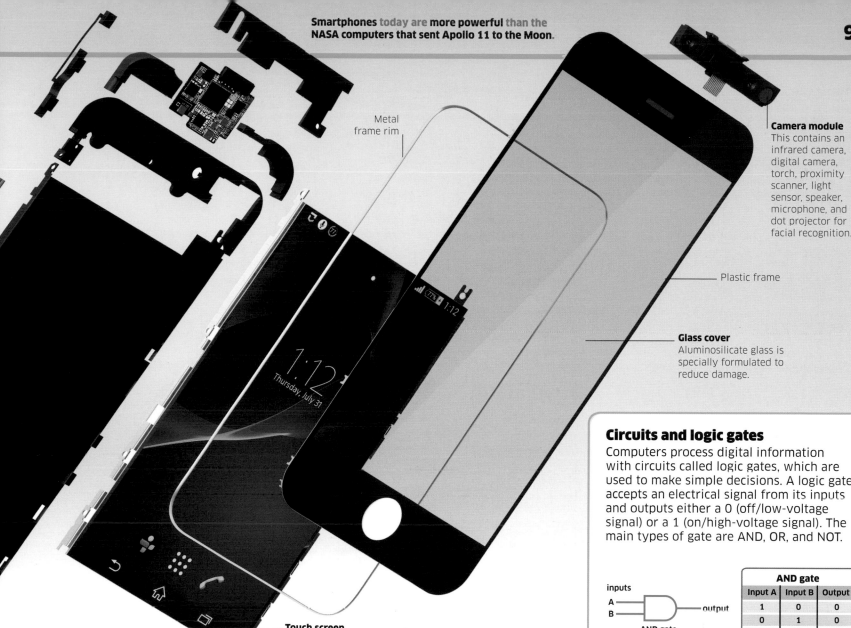

Metal frame rim

Camera module
This contains an infrared camera, digital camera, torch, proximity scanner, light sensor, speaker, microphone, and dot projector for facial recognition.

Plastic frame

Glass cover
Aluminosilicate glass is specially formulated to reduce damage.

Touch screen
A grid of sensors registers touch as electrical signals, which are sent to the processor. This interprets the gesture and relates it to the app being run.

Circuits and logic gates

Computers process digital information with circuits called logic gates, which are used to make simple decisions. A logic gate accepts an electrical signal from its inputs and outputs either a 0 (off/low-voltage signal) or a 1 (on/high-voltage signal). The main types of gate are AND, OR, and NOT.

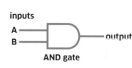

inputs
A
B
output
AND gate

AND gate		
Input A	Input B	Output
1	0	0
0	1	0
0	0	0
1	1	1

AND gate
This compares two numbers and switches on only if both the numbers are 1. There will only be an output if both inputs are on.

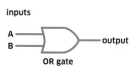

inputs
A
B
output
OR gate

OR gate		
Input A	Input B	Output
0	0	0
0	1	1
1	0	1
1	1	1

OR gate
This switches on if either of two numbers is 1. If both numbers are 0, it switches off. There will be an output if one or both inputs are on.

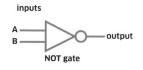

inputs
A
B
output
NOT gate

NOT gate	
Input	Output
0	1
1	0

NOT gate
This reverses (inverts) whatever goes into it. A 0 becomes a 1, and vice versa. The output is on only if the input is off. If the input is on, the output is off.

Digital electronics

Most technology we use today is digital. Our devices convert information into numbers or digits and process these numbers in place of the original information. Digital cameras turn images into patterns of numbers, while mobile phones send and receive calls with signals representing strings of numbers. These are sent in a code called binary, using only the numerals 1 and 0 (rather than decimal, 0–9).

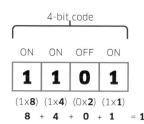

4-bit code

ON	ON	OFF	ON
1	**1**	**0**	**1**

(1x**8**) (1x**4**) (0x**2**) (1x**1**)
8 + 4 + 0 + 1 = 13

Binary numbers
In binary, the position of 1s and 0s corresponds to a decimal value. Each binary position doubles in decimal value from right to left (1, 2, 4, 8) and these values are either turned on (x1) or off (x0). In the 4-bit code shown, the values of 8, 4, and 1 are all "on", and when added together equal 13.

Analogue to digital

A sound wave made by a musical instrument is known as analogue information. The wave rises and falls as the sound rises and falls. A wave can be measured at different points to produce a digital version with a pattern more like a series of steps than a wave form.

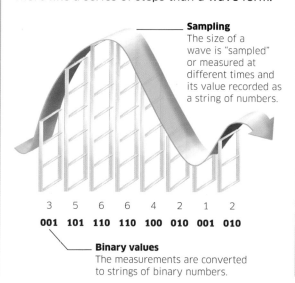

Sampling
The size of a wave is "sampled" or measured at different times and its value recorded as a string of numbers.

3	5	6	6	4	2	1	2
001	**101**	**110**	**110**	**100**	**010**	**001**	**010**

Binary values
The measurements are converted to strings of binary numbers.

FORCES

Invisible forces are constantly at play in our day-to-day life, from the wind rustling the leaves of trees to the tension in the cables of a suspension bridge. A force is any push or pull. Forces can change an object's speed or direction of motion, or can change its shape. English scientist Isaac Newton worked out how forces affect motion over three hundred years ago (see pp.98–99). His principles are still applied in many fields of science, engineering, and in daily life today.

WHAT IS A FORCE?

A force can be a push or a pull. Although you can't see a force, you can often see what it does. A force can change the speed, direction, or shape of an object. Motion is caused by forces, but forces don't always make things move – balanced forces are essential for building stability.

Contact forces

When one object comes into contact with another and exerts a force, this is called a contact force. Either a push or a pull, this force changes the direction, speed, or shape of the object.

Changing direction
if a player bounces a football against a wall during practice, the wall exerts a force on the ball that changes its direction.

Changing speed
If a player kicks, back-heels, or volleys a football, the force that is applied changes the ball's speed.

Changing shape
Kicking or stepping on the football applies a force that momentarily squashes it, changing the ball's shape.

Non-contact forces

All forces are invisible, but some are exerted without physical contact between objects. The closer two objects are to each other, the stronger is the force.

Gravity
Gravity is a force of attraction between objects with mass. Every object in the Universe pulls on every other object.

Magnetism
A magnet creates a magnetic field around it. If a magnetic material is brought into the field, a force is exerted on it.

Static electricity
A charged object creates an electric field. If another charged object is moved into the field, a force acts on it.

Weight, gravity, and mass

Weight is not the same as mass, which is a measure of how much matter is in an object. Weight is the force acting on that matter and is the result of gravity. The mass of an object is the same everywhere, but its weight can change.

Measuring forces
Forces can be measured using a force meter, which contains a spring connected to a metal hook. The spring stretches when a force is applied to the hook. The bigger the force, the longer the spring stretches and the bigger the reading. The unit of force is the newton (N).

Calculating weight
Mass is measured in kilograms (kg). Weight can be calculated as mass x gravity (N/kg). The pull of gravity at Earth's surface is roughly 10 N/kg, so an object with a mass of 1 kg weighs 10 N.

BALANCED AND UNBALANCED FORCES

Not all forces acting on an object make it move faster or in a different direction: forces on a bridge must be balanced for the structure to remain stable. In a tug of war, there's no winner while the forces are balanced; it takes a greater force from one team to win.

Balanced forces
If two forces acting on an object are equal in size but opposite in direction, they are balanced. An object that is not moving will stay still, and an object in motion will keep moving at the same speed in the same direction.

The tension in the rope is 500 N.

250 N   250 N

Unbalanced forces
If two forces acting on an object are not equal, they are unbalanced. An object that is not moving will start moving, and an object in motion will change speed or direction.

150 N   350 N

DEFORMING FORCES

When a force acts on an object that cannot move, or when a number of different forces act in different directions, the whole object changes shape. The type of distortion an object undergoes depends on the number, directions, and strengths of the forces acting upon it, and on its structure and composition – if it is elastic (returns to its original shape) or plastic (deforms easily but does not return to its original shape). Brittle materials fracture, creep, or show fatigue if forces are applied to them.

Compression
When two or more forces act in opposite directions and meet in an object, it compresses and bulges.

Tension
When two or more forces act in opposite directions and pull away from an elastic object, it stretches.

Torsion
Turning forces, or torques, that act in opposite directions twist the object.

Bending
When several forces act on an object in different places, the object bends (if malleable) or snaps.

Resultant forces
A force is balanced when another force of the same strength is acting in the opposite direction. Overall, this has the same effect as no force at all.

RESULTANT FORCE: 0 N

When opposing teams pull with equal force, the resultant force is 0 N.

RESULTANT FORCE: 200 N

One team pulls with more force than the other. The resultant force is 200 N.

TURNING FORCES

Instead of just moving or accelerating an object in a line, or sending an object off in a straight line in a different direction, forces can also be used to turn an object around a point known as an axis or a pivot. This kind of force works on wheels, see-saws, and fairground rides such as carousels. The principles behind these turning forces are also used in simple machines (see pp.106–107).

Moment

When a force acts to turn an object around a pivot, the effect of the force is called its moment. The turning effect of a force depends on the size of the force and how far away from the pivot the force is acting. Calculated as force (N) x distance (m), moment is measured in newton metres (Nm).

Sitting closer to the pivot of a see-saw increases the moment.

A greater weight increases the moment.

The centre of a see-saw is its pivot.

Centripetal forces

A constant force has to be applied to keep an object turning in a circle, obeying Newton's first law of motion (see pp.98–99). Known as centripetal force, it pulls the turning object towards the centre of rotation – imagine a yo yo revolving in a circle on its string – continually changing its direction, while the motion changes its speed. Without this force, the object would move in a straight line away from the centre.

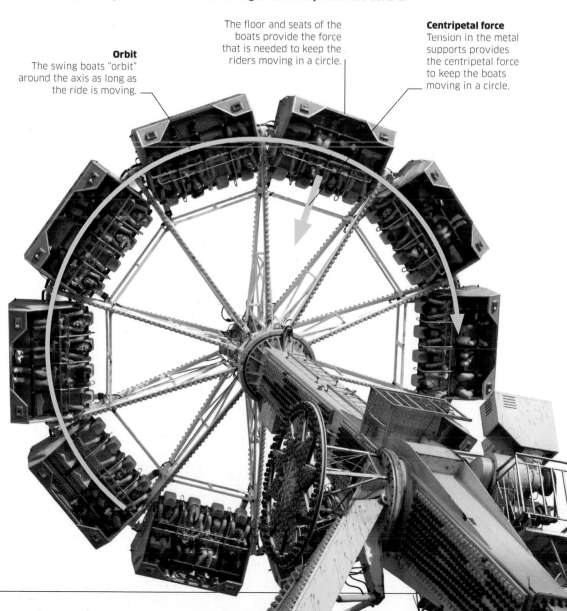

Orbit
The swing boats "orbit" around the axis as long as the ride is moving.

The floor and seats of the boats provide the force that is needed to keep the riders moving in a circle.

Centripetal force
Tension in the metal supports provides the centripetal force to keep the boats moving in a circle.

In frictionless space, **spacecraft travel at a constant velocity** – obeying Newton's first law of motion.

Laws of motion

When a force acts on an object that is free to move, the object will move in accordance with Newton's three laws of motion.

English physicist and mathematician Isaac Newton published his laws of motion in 1687. They explain how objects move – or don't move – and how they react with other objects and forces. These three scientific laws form the basis of what is known as classical mechanics. Modern physics shows that Newton's laws are not perfectly accurate, but they are still useful in everyday situations.

Ariane 5

The Ariane 5 rocket is a launch vehicle used to deliver massive payloads, such as communication satellites, into orbit. Causing a rocket to accelerate upwards requires enormous forces to overcome the gravity pulling it downwards. Hot gases expand, exerting forces on the walls of the combustion chamber to lift the rocket. The walls of the chamber produce a reaction force that pushes back on the gases, which escape at high speed through the nozzles at the bottom of the engine. These forces create acceleration.

Fairing protects satellites during lift-off.

THRUST

NET FORCE

A global navigation system is carried into orbit.

A communication satellite is mounted in the upper stage.

Vehicle equipment bay
The rocket's "brain", this contains equipment that guides and tracks the rocket.

Upper cryogenic stage
The upper stage is powered by a separate engine to position satellites in orbit.

Liquid oxygen tank contains 150 tonnes of oxidizer.

Main cryogenic stage
The main stage holds liquid hydrogen that mixes with liquid oxygen in the combustion chamber to combust.

Liquid hydrogen tank contains 25 tonnes of fuel.

First law of motion

Any object will remain at rest, or move in a straight line at a steady speed, unless an external force acts upon it. So, a football is stationary until it is kicked and then moves until other forces stop it. This is known as inertia. If all external forces are balanced, the object will maintain a constant velocity. For an object that is not moving, this is zero.

At rest
Gravity acts on the football, but the ground stops it from moving so it remains at rest.

Force causes motion
The impact of a boot kicking the football applies a force that accelerates the ball.

Force stops motion
The football slows down due to friction and stops when it meets a boot.

Second law of motion

When a force acts on an object, the object will generally move in the direction of the force. This causes a change in velocity, known as acceleration. The larger the force, the greater an object's acceleration will be. The more massive an object is, the greater the force needed to accelerate it. This is written as force = mass x acceleration.

Force

Acceleration

Small mass, small force
A force causes an object to accelerate, changing its velocity per second, at a certain rate.

Small mass, double force
If the mass stays the same but the force doubles, the object will accelerate at twice the rate.

Double mass, double force
If the mass doubles and the force doubles again, the rate of acceleration stays the same.

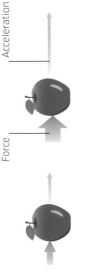

773 tonnes – the weight of Ariane 5 at lift-off. Its lift-off thrust is 1,340 tonnes.

To reach a **low Earth orbit**, a rocket must generate **enough thrust** to reach **a speed of 29,000 km/h (18,000 mph)**. **99**

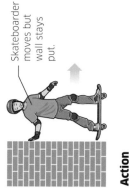

Solid rocket booster
Two solid-propellant boosters deliver more than 90 per cent of thrust in a short blast at lift-off.

Vulcain 2 engine
The main engine burns for 10 minutes to provide thrust.

Fuel combusts in the combustion chamber, generating hot gases at high pressure.

Rocket nozzle expels hot exhaust gases, which push backwards.

Newton's three laws at work

A rocket taking off shows Newton's laws in action. Before lift-off, the rocket's enormous weight – the result of gravity pulling it down towards Earth – is balanced by the upwards force of the launch pad so it remains stationary (first law). Fuel combustion in the engines creates thrust, propelling the rocket forwards (second law). The thrust that pushes the rocket forwards is a reaction to the hot exhaust gases pushing backwards (third law).

WEIGHT

Velocity, speed, and acceleration

Speed is a measure of the rate at which a distance is covered. Velocity is not the same as speed; it measures direction as well as speed of movement. Acceleration measures the rate of change of velocity. Speeding up, turning, and slowing down are all acceleration.

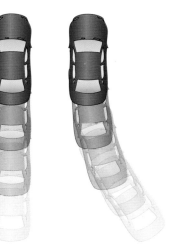

Increasing speed
When a force is applied to an object, its speed increases – it accelerates.

Changing direction
When an object changes direction, its velocity changes – this is also a type of acceleration.

Decreasing speed
When a force slows a moving object down, its speed decreases – it decelerates, or accelerates negatively.

Third law of motion

Forces come in pairs, and any object will react to a force applied to it. The force of reaction is equal and acts in an opposite direction to the force that produces it. If one object is immobile, then the other will move. If both objects can move, then the object with less mass will accelerate more than the other. Every action has an equal and opposite reaction.

Skateboarder moves but wall stays put.

Skateboarders move in opposite directions at same velocity.

Action
If a skateboarder pushes a wall, the wall pushes back with a reaction force that causes the boarder to roll away from it.

Reaction
If one skateboarder pushes another, action and reaction cause both boarders to roll away from each other.

Momentum

A moving object carries on moving because it has momentum. It will keep moving until a force stops it. However, when it collides with another object, momentum will be transferred to the second object.

Newton's cradle
Energy is conserved when the balls collide.

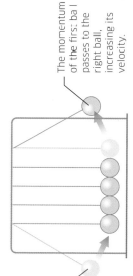

As the left ball hits the line of other balls, its velocity decreases and its momentum falls to zero.

The momentum of the first ball passes to the right ball, increasing its velocity.

Relative velocity

The velocity of an object is its speed in a particular direction. Two objects travelling at the same speed but in opposite directions, or at different speeds in the same direction, have different velocities.

Same direction, same speed
The relative velocity of the two cars is 0km/h (0mph).

CAR TRAVELLING AT 50KM/H (30MPH)

CAR TRAVELLING AT 50KM/H (30MPH)

Same direction, different speeds
The relative velocity of the two cars is 15km/h (10mph).

CAR TRAVELLING AT 65KM/H (40MPH)

CAR TRAVELLING AT 50KM/H (30MPH)

Opposite direction, same speed
The relative velocity of cars on a collision course is 100km/h (60mph).

CAR TRAVELLING AT 50KM/H (30MPH)

CAR TRAVELLING AT 50KM/H (30MPH)

Friction

Friction is a force that occurs when a solid object rubs against or slides past another, or when it moves through a liquid or a gas. It always acts against the direction of movement.

The rougher surfaces are and the harder they press together, the stronger the friction – but friction occurs even between very smooth surfaces. Friction can be useful – it helps us to stand, walk, and run – but it can also be a hindrance, slowing movement and making machines inefficient. A by-product of friction is heat.

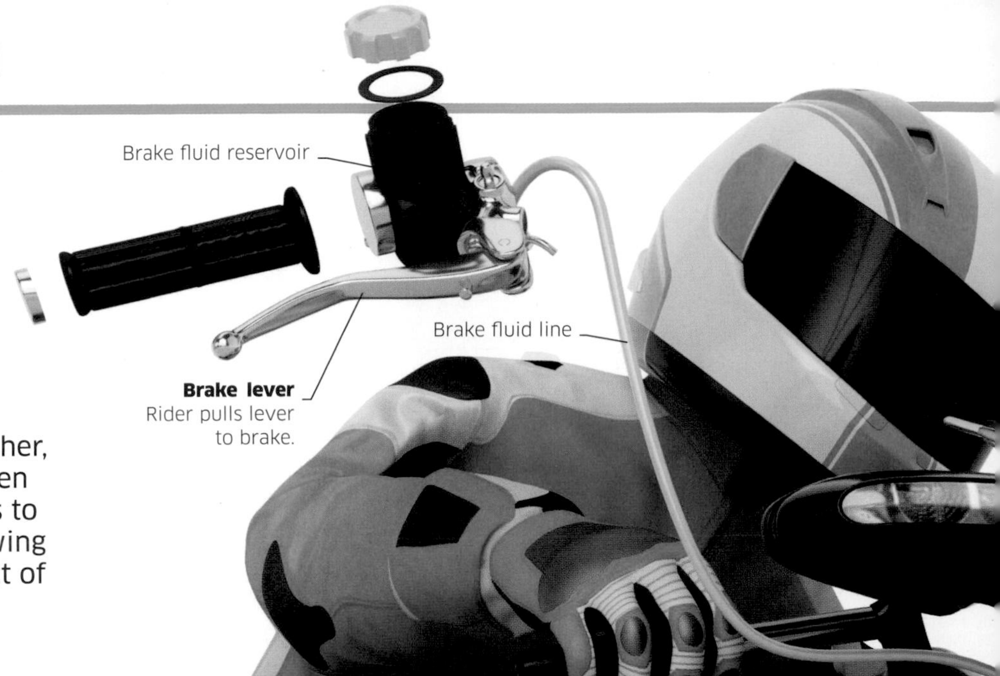

Brake fluid reservoir

Brake fluid line

Brake lever
Rider pulls lever to brake.

Leathers
Leather clothing protects the rider from friction burns and grazes in the event of an accident.

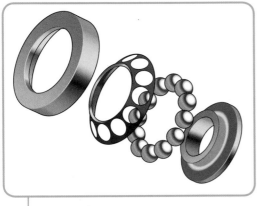

Ball bearings
Inside the axle of a wheel, ball bearings reduce friction between the turning parts. The balls rotate as the wheel turns, making the surfaces slide more easily. They are lubricated with oil.

Tyre tread
The tread – the pattern of grooves on the tyre – helps to maintain grip on different types of surface.

Brake pedal
Friction between the foot and pedal maintains grip.

Fairings
On the side of the bike, fairings reduce drag.

Fish and aquatic mammals such as whales and dolphins have streamlined body shapes to reduce water resistance.

If the re-entry angle of a spacecraft is too steep, the braking effect due to atmospheric friction will cause the spacecraft to break up.

101

Friction in a motorbike

The force of friction both helps and hinders a motorbike rider. Friction between the tyres and ground is essential for movement and grip, and is the force behind braking. Drag, the friction that occurs between air and the bike, slows the rider down, and friction between moving parts makes the bike less efficient.

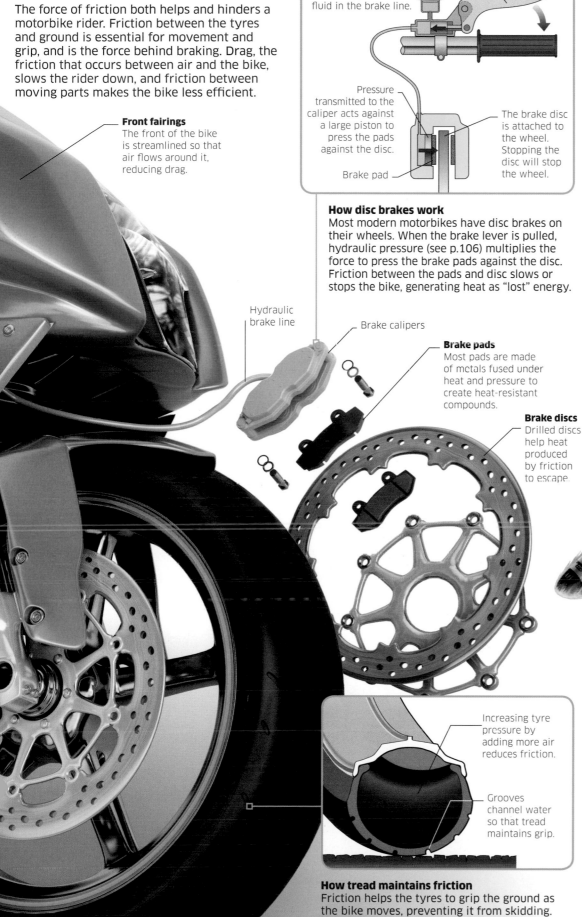

Front fairings
The front of the bike is streamlined so that air flows around it, reducing drag.

Pulling a lever pushes a small piston, exerting pressure on fluid in the brake line.

Pressure transmitted to the caliper acts against a large piston to press the pads against the disc.

Brake pad

The brake disc is attached to the wheel. Stopping the disc will stop the wheel.

How disc brakes work

Most modern motorbikes have disc brakes on their wheels. When the brake lever is pulled, hydraulic pressure (see p.106) multiplies the force to press the brake pads against the disc. Friction between the pads and disc slows or stops the bike, generating heat as "lost" energy.

Hydraulic brake line

Brake calipers

Brake pads
Most pads are made of metals fused under heat and pressure to create heat-resistant compounds.

Brake discs
Drilled discs help heat produced by friction to escape.

Increasing tyre pressure by adding more air reduces friction.

Grooves channel water so that tread maintains grip.

How tread maintains friction

Friction helps the tyres to grip the ground as the bike moves, preventing it from skidding. The tread is designed to channel water through grooves, so that the tyres still grip on wet and muddy roads.

Fluid resistance (drag)

When an object moves through a fluid, it pushes the fluid aside. That requires energy, so the object slows down – or has to be pushed harder; this is known as form drag. Fluid also creates friction as it flows past the object's surface; this is called skin friction.

Water resistance
When a boat moves through water, it pushes water out of the way. The water resists, rising up as bow and stern waves and creating transverse waves in the boat's wake.

Air resistance

When an object moves through air, the drag is called air resistance. The bigger and less streamlined the object and the faster the object is moving, the greater the drag. When spacecraft re-enter the atmosphere, moving very fast, the drag heats their surfaces to as much as 1,500°C (2,750°F).

Helpful and unhelpful friction

It is tempting to think of friction as an unhelpful force that slows movement, but friction can be helpful too. Without friction between surfaces, there would be no grip and it would be impossible to walk, run, or cycle. However, the boot is on the other foot for skiers, snowboarders, and skaters, who minimize friction to slide.

Reducing friction
The steel blades of ice skates reduce friction, enabling skaters to glide across ice.

Increasing friction
The treads of rubber-soled mountain boots increase friction and grip for climbers.

102 energy and forces ∘ **GRAVITY**

23 kg (51 lbs) – the **weight on Mars**, due to lower gravity, of a person weighing **62 kg** (137 lbs) on Earth.

Law of Falling Bodies

Gravity pulls more strongly on heavier objects – but heavier objects need more force to make them speed up than lighter ones. Galileo was the first person to realize, in 1590, that any two objects dropped together should speed up at the same rate and hit the ground together. We are used to lighter objects falling more slowly – because air resistance slows them more.

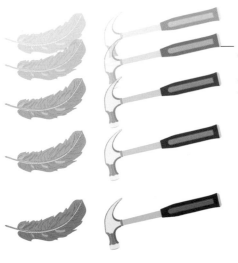

In the near-frictionless environment of the Moon, a heavy hammer and a light feather fall at the same rate.

Falling in a vacuum
In 1971, astronaut Dave Scott proved Galileo right when he dropped a feather and a hammer on the Moon.

Law of Universal Gravitation

In 1687, English scientist Isaac Newton came up with his Law of Universal Gravitation. It states that any two objects attract each other with a force that depends on the masses of the objects and the distance between them.

Equal and opposite
The gravitational force between two objects pulls equally on both of them – whatever their relative mass – but in opposite directions.

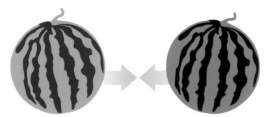

Double the mass
If one object's mass is doubled, the gravitational force doubles. If the mass of both objects is doubled (as here), the force is four times as strong.

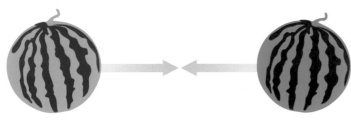

Double the distance
If the distance between two objects is doubled, the gravitational force is quartered.

Gravity and orbits

Newton used his understanding of gravity (see left) and motion to work out how planets, including Earth, remain in their orbits around the Sun. He realized that without gravity Earth would travel in a straight line through space. The force of gravity pulls Earth towards the Sun, keeping it in its orbit. Earth is constantly falling towards the Sun, but never gets any closer. If Earth slowed down or stopped moving, it would fall into the Sun!

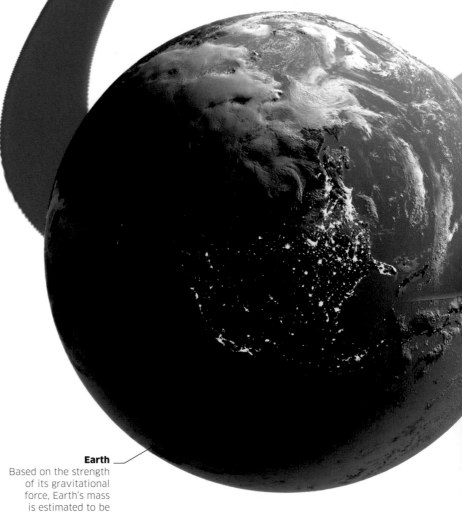

Elliptical orbit
Earth's orbit around the Sun is in fact elliptical (a squashed circle), not circular.

Speed of travel
If Earth was not speeding through space, gravity would pull it into the Sun.

Earth
Based on the strength of its gravitational force, Earth's mass is estimated to be 5.9 sextillion (that's 5.9 with 12 zeroes) tonnes!

The force of gravity 100 km (62 miles) above Earth is
3 per cent less
than at sea level on Earth.

9.8 m/s² – the **rate** at which a falling object **accelerates towards Earth**.

Gravity is the **weakest of the four fundamental forces**: strong nuclear, electromagnetic, weak nuclear, and gravitational.

103

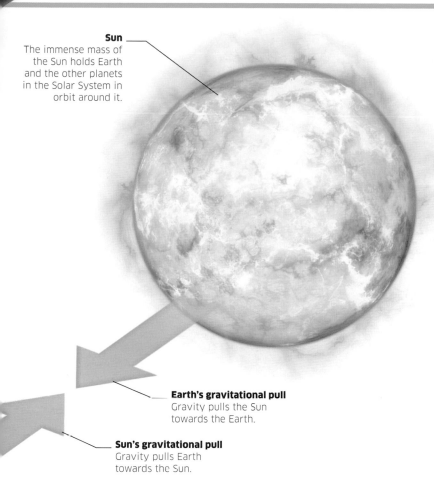

Sun
The immense mass of the Sun holds Earth and the other planets in the Solar System in orbit around it.

Gravity

Gravity is a force of attraction between two objects. The more mass the objects have and the closer they are to each other, the greater the force of attraction.

Earth's gravity is the gravitational force felt most strongly on the planet: it is what keeps us on the ground and stops us floating off into space. In fact, we pull on Earth as much as Earth pulls on us. Gravity also keeps the planets in orbit around the Sun, and the Moon around Earth. Without it, each planet would travel in a straight line off into space.

The best way scientists can explain gravity is with the General Theory of Relativity, formulated by Albert Einstein in 1915. According to this theory, gravity is actually caused by space being distorted around objects with mass. As objects travel through the distorted space, they change direction. So, according to Einstein, gravity is not a force at all!

Earth's gravitational pull
Gravity pulls the Sun towards the Earth.

Sun's gravitational pull
Gravity pulls Earth towards the Sun.

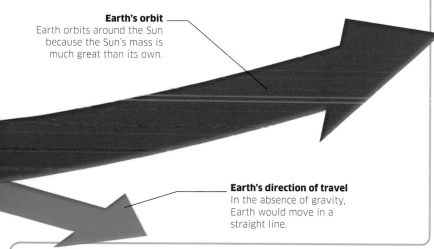

Earth's orbit
Earth orbits around the Sun because the Sun's mass is much great than its own.

Earth's direction of travel
In the absence of gravity, Earth would move in a straight line.

Mass and weight

Mass is the amount of matter an object contains, which stays the same wherever it is. It is measured in kilograms (kg). Weight is a force caused by gravity. The more mass an object has and the stronger the gravity, the greater its weight. Weight is measured in newtons (N).

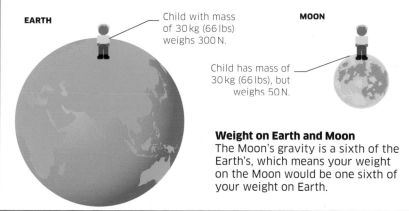

EARTH

Child with mass of 30 kg (66 lbs) weighs 300 N.

MOON

Child has mass of 30 kg (66 lbs), but weighs 50 N.

Weight on Earth and Moon
The Moon's gravity is a sixth of the Earth's, which means your weight on the Moon would be one sixth of your weight on Earth.

Tides

The gravitational pull of the Moon and the Sun cause the oceans to bulge outwards. The Moon's pull on the oceans is strongest, because it is closest to Earth, and it is the main cause of the tides. However, at certain times of each lunar month, the Sun's gravity also plays a role, increasing or decreasing the height of the tides.

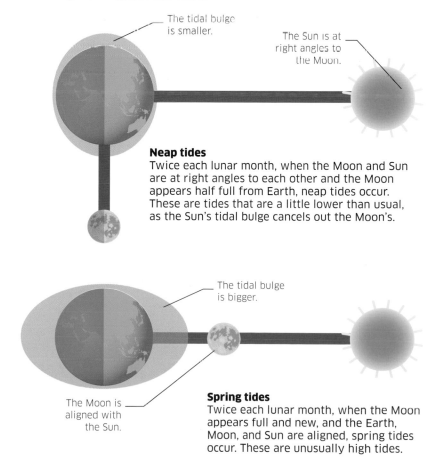

The tidal bulge is smaller.

The Sun is at right angles to the Moon.

Neap tides
Twice each lunar month, when the Moon and Sun are at right angles to each other and the Moon appears half full from Earth, neap tides occur. These are tides that are a little lower than usual, as the Sun's tidal bulge cancels out the Moon's.

The tidal bulge is bigger.

The Moon is aligned with the Sun.

Spring tides
Twice each lunar month, when the Moon appears full and new, and the Earth, Moon, and Sun are aligned, spring tides occur. These are unusually high tides.

Pressure

Pressure is the push on a surface created by one or more forces. How much pressure is exerted depends upon the strength of the forces and the area of the surface. Walk over snow in showshoes and you won't sink in – but walk on grass in stiletto heels and you will.

Solids, liquids, and gases can apply pressure onto a surface because of their weight pressing down on it. The pressure applied by liquids and gases can be increased by squashing them. Pressure is measured in newtons per square metre (N/m²) – also called Pascals (Pa) – or pounds per square inch (psi).

Atmospheric and water pressure

Near sea level, the weight of the air around us presses with a force of about 100,000 Pa (15 psi). Pressure decreases with altitude, because there is less air above pressing down. In the ocean, pressure increases quickly with depth, since water is denser than air.

400 km (250 miles)
As a Soyuz spacecraft travels to the International Space Station (ISS), which orbits at 400 km (250 miles), gas molecules are so few and far between that air pressure is almost non-existent. The space station's atmosphere is maintained at the same pressure as sea level.

35,000 m (115,000 ft)
As weather balloons ascend into the stratosphere, they expand from 2 m (6 ft 6 in) to 8 m (26 ft) across as air pressure decreases to just 1,000 Pa (0.1 psi). The gas molecules within the balloon spread out as pressure from outside diminishes.

18,000 m (60,000 ft)
Above this altitude (the Armstrong limit) humans cannot survive in an unpressurized environment. Air pressure is 7,000 Pa (1 psi) and exposed body fluids such as saliva and moisture in the lungs will boil away – but not blood in the circulatory system.

11,000 m (36,000 ft)
This is the cruising altitude of passenger jets. As a plane lifts off, your ears may pop due to the change in pressure: air trapped in the inner ear stays at the same pressure, but air pressure outside changes, exerting a force on your eardrum. Pressure falls to 23,000 Pa (3 psi) on the plane's exterior.

8,848 m (28,871 ft)
At Everest's summit, atmospheric pressure is one third of that at sea level: 33,000 Pa (4.5 psi). It is hard to make tea as water boils at 72°C (162°F) – not hot enough for a good brew. Liquids boil when the particles of which they are made move fast enough to have the same pressure as air – so when pressure falls, the boiling point is lower.

Felix Baumgartner makes a record-breaking skydive from 38,964 m (127,852 ft).

Pilots of fighter jets cruising at 15,000 m (50,000 ft) wear pressure suits.

40,000 m (130,000 ft)
ABOVE SEA LEVEL

35,000 m (115,000 ft)

40,000 m
(130,000 ft)

30,000 m (98,000 ft)

25,000 m (82,000 ft)

20,000 m (65,500 ft)

The **record altitude for a jet plane** with a pressurized cockpit **is 37,649 m (123,520 ft)**, set by a Russian MiG-25M.

The **record depth for a scuba dive**, set by Egyptian diver Ahmed Gabr is **332.5 m (1,090 ft) below sea level.**

105

5,500 m (18,000 ft)
One half of the atmosphere is contained between Earth's surface and 5,500 m (18,000 ft), where air pressure is 50,000 Pa (7.3 psi). The other half is between this altitude and 30,000 m (100,000 ft).

5,300 m (17,400 ft)
At Everest base camp, atmospheric pressure is about half that at sea level. Altitude sickness is common as air pressure falls to 51,000 Pa (7.4 psi) and there is a low concentration of gas molecules. Climbers pause here to acclimatize and few go higher without extra oxygen.

1,500 m (5,000 ft)
Air pressure decreases to 84,000 Pa (12 psi) at this altitude and breathing is difficult. Lower air density means there are fewer molecules in the same volume of air so people have to breathe faster and deeper to take in the same amount of oxygen.

0 m (0 ft)
At sea level, the pressure pushing down on the surface, known as "one atmosphere", is 101,000 Pa (15 psi). It is the result of the weight of all the air above that surface.

−9.75 m (−32 ft)
Atmospheric pressure is double that at sea level: 200,000 Pa (30 psi). This means that a 9.75-m (32-ft) column of water weighs as much as the entire column of air above it from outer space to 0 m (0 ft).

−40 m (−130 ft)
The normal depth limit for a qualified scuba diver. Pressure here is 500,000 Pa (73 psi) – nearly five times sea level. The spongy tissue of the lungs begins to contract, making it hard to breathe. Diving tanks contain compressed, oxygen-enriched air to overcome this.

−4,000 m (−13,000 ft)
The average depth of the oceans is six times that of the maximum crush depth of most modern submarines, which can survive pressure of 40 million Pa (5,800 psi) – four hundred times what it is at sea level.

−10,994 m (−36,070 ft)
The *Deepsea Challenger* submersible dove close to the deepest known point in Earth's ocean, Challenger Deep in the Marianas Trench, where pressure is 110 million Pa (16,040 psi) – more than a thousand times atmospheric pressure at sea level.

Skydivers typically jump from 3,500 m (11,500 ft).

Herbert Nitsch makes a record-breaking free dive to −214 m (−702 ft).

Russian submarine *Komsomolets K-278* dives to −1,020 m (−3,346 ft).

10,000 m (33,000 ft)

5,000 m (16,500 ft)

0 m (0 ft)

−5,000 m (−16,500 ft)

−4,000 m (−13,000 ft)

−11,000 m (−36,000 ft)
BELOW SEA LEVEL

106 energy and forces ○ **SIMPLE MACHINES**

A complex system of levers connects the keys of a piano to the hammers that hit the strings to **sound the notes**.

Simple machines

A machine is anything that changes the size or direction of a force, making work easier. Simple machines include ramps, wedges, screws, levers, wheels, and pulleys.

Complex machines such as cranes and diggers combine a number of simple machines, but whatever the scale, the physical principles remain the same. Many of the most effective machines are the simplest – a sloping path (ramp), a knife (wedge), a jar lid (screw), scissors, nutcrackers, and tweezers (levers), a tap (wheel and axle), or hoist (pulley), for example. Hydraulics and pneumatics use the pressure in fluids (liquids and gases) to transmit force.

Lever
The crane's boom is a long, third-class lever. When a hydraulic ram applies a force greater than the load between the load and the fulcrum, the crane lifts the load.

Applying effort halfway up the boom doubles the distance the load travels.

LOAD

EFFORT

FULCRUM

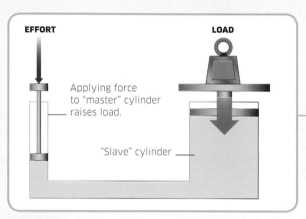

EFFORT

LOAD

Applying force to "master" cylinder raises load.

"Slave" cylinder

Hydraulics
A hydraulic system makes use of pressure in a liquid by applying force (effort) to a "master" cylinder, which increases fluid pressure in a "slave" cylinder. The hydraulic ram lifts the crane's boom by using pressure from fluid in the cylinder to push a piston.

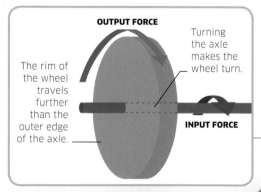

OUTPUT FORCE

Turning the axle makes the wheel turn.

The rim of the wheel travels further than the outer edge of the axle.

INPUT FORCE

Wheel and axle
A wheel with an axle can be used in two ways: either by applying a force to the axle to turn the wheel, which multiplies the distance travelled; or by applying a force to the wheel to turn the axle, like a spanner.

Ramp and wedge
Also known as an inclined plane, a ramp reduces the force needed to move an object from a lower to a higher place. A wedge acts like a moving inclined plane, applying a greater force to raise an object.

LOAD

EFFORT

Less effort is needed to push a load up a ramp, but the load has to move further along the slope than it moves vertically.

The tool known as **Archimedes' screw pump** has been used to **transport water for irrigation** since the **7th century BCE**.

107

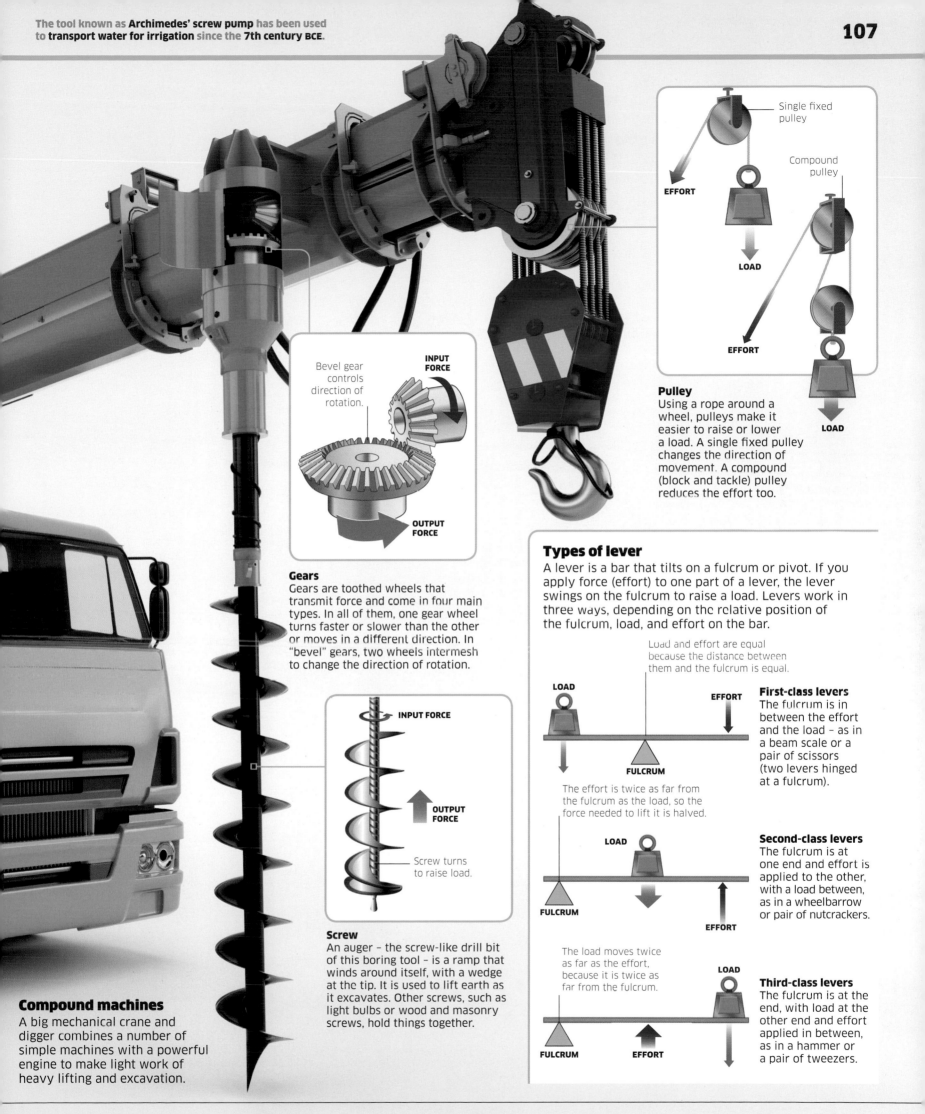

Single fixed pulley

Compound pulley

EFFORT

LOAD

EFFORT

LOAD

LOAD

Pulley

Using a rope around a wheel, pulleys make it easier to raise or lower a load. A single fixed pulley changes the direction of movement. A compound (block and tackle) pulley reduces the effort too.

Bevel gear controls direction of rotation.

INPUT FORCE

OUTPUT FORCE

Gears

Gears are toothed wheels that transmit force and come in four main types. In all of them, one gear wheel turns faster or slower than the other or moves in a different direction. In "bevel" gears, two wheels intermesh to change the direction of rotation.

Types of lever

A lever is a bar that tilts on a fulcrum or pivot. If you apply force (effort) to one part of a lever, the lever swings on the fulcrum to raise a load. Levers work in three ways, depending on the relative position of the fulcrum, load, and effort on the bar.

Load and effort are equal because the distance between them and the fulcrum is equal.

LOAD

EFFORT

FULCRUM

First-class levers
The fulcrum is in between the effort and the load – as in a beam scale or a pair of scissors (two levers hinged at a fulcrum).

The effort is twice as far from the fulcrum as the load, so the force needed to lift it is halved.

LOAD

FULCRUM

EFFORT

Second-class levers
The fulcrum is at one end and effort is applied to the other, with a load between, as in a wheelbarrow or pair of nutcrackers.

INPUT FORCE

OUTPUT FORCE

Screw turns to raise load.

Screw

An auger – the screw-like drill bit of this boring tool – is a ramp that winds around itself, with a wedge at the tip. It is used to lift earth as it excavates. Other screws, such as light bulbs or wood and masonry screws, hold things together.

The load moves twice as far as the effort, because it is twice as far from the fulcrum.

LOAD

FULCRUM

EFFORT

Third-class levers
The fulcrum is at the end, with load at the other end and effort applied in between, as in a hammer or a pair of tweezers.

Compound machines

A big mechanical crane and digger combines a number of simple machines with a powerful engine to make light work of heavy lifting and excavation.

Floating

Why does an apple float but a gold apple of the same size sink? How do ships carrying a cargo across the sea stay afloat? And what makes a balloon float in air?

Fluids (liquids and gases) exert pressure on the surface of any object immersed in them. Pressure in a fluid increases with depth, so the pressure pushing upwards on the bottom of an object is greater than the pressure pushing downwards on the top. This results in an upward force, called "upthrust". If the upthrust on an object is greater than, or equal to, its weight, the object floats. If the upthrust is less than its weight, the object sinks. Same-sized objects of different densities weigh more or less, so one object may float while another of the same size sinks.

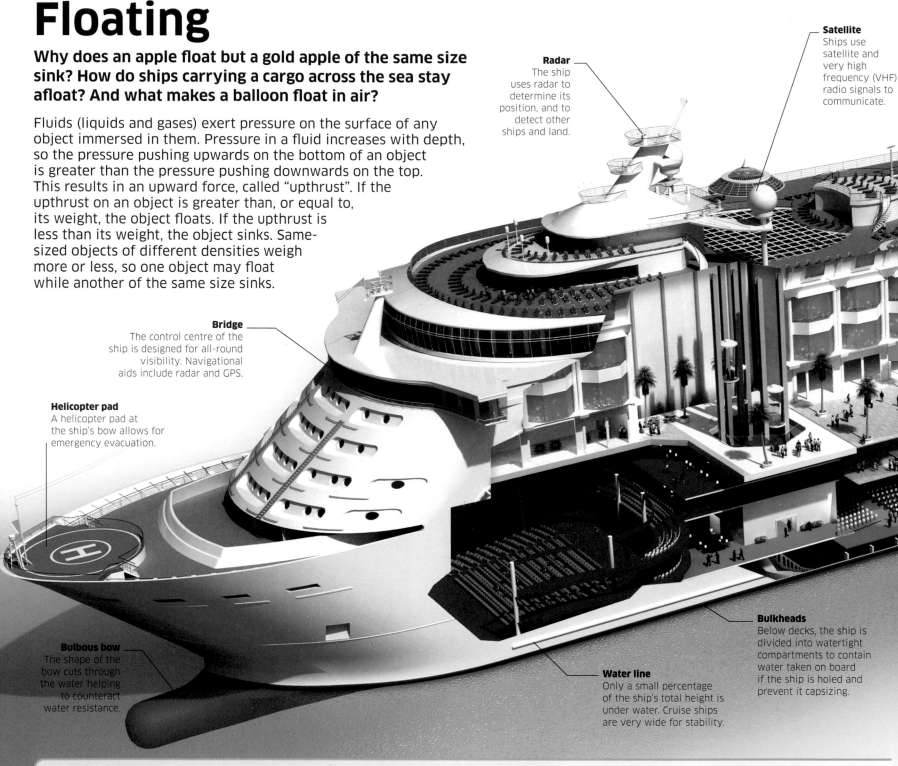

Radar
The ship uses radar to determine its position, and to detect other ships and land.

Satellite
Ships use satellite and very high frequency (VHF) radio signals to communicate.

Bridge
The control centre of the ship is designed for all-round visibility. Navigational aids include radar and GPS.

Helicopter pad
A helicopter pad at the ship's bow allows for emergency evacuation.

Bulbous bow
The shape of the bow cuts through the water helping to counteract water resistance.

Water line
Only a small percentage of the ship's total height is under water. Cruise ships are very wide for stability.

Bulkheads
Below decks, the ship is divided into watertight compartments to contain water taken on board if the ship is holed and prevent it capsizing.

Water density

When ocean trade routes opened up around the globe, sailors were surprised to find their carefully loaded ships sank when they got near the equator. This was because the density of warm tropical waters was less than that of cool northern waters, and so provided less upthrust. When the ships entered freshwater ports, the water density was lower still, and ships were even more likely to sink.

TROPICAL FRESH WATER

Warm water that is not salt has a low density, so a ship floats low in it.

TROPICAL SEAWATER

Salt water has a higher density than fresh water, so a ship floats higher.

SUMMER TEMPERATE SEAWATER

As salt water cools, its density increases and a ship becomes more buoyant.

WINTER NORTH ATLANTIC SEAWATER

In the freezing cold waters of the North Atlantic, ships float high in the water.

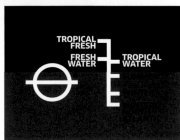

The Plimsoll line
On a ship's hull, this mark shows the depth to which the ship may be immersed when loaded. This varies with a ship's size, type of cargo, time of year, and the water densities in port and at sea.

Greek philosopher **Archimedes first established the principle of buoyancy, or how things float**, in the **2nd century BCE**.

109

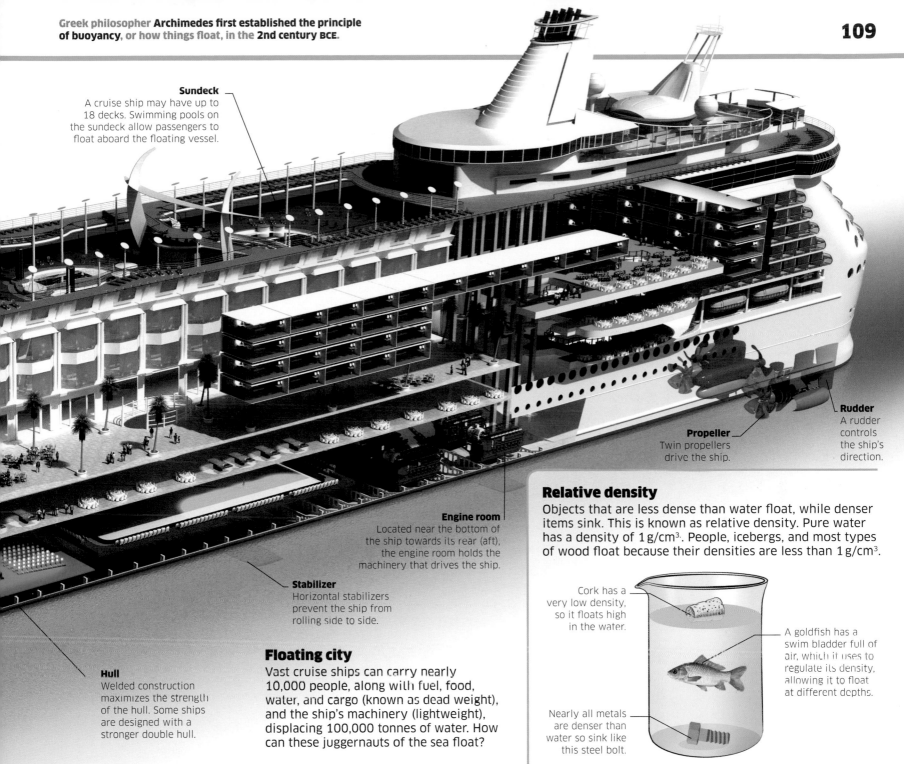

Sundeck
A cruise ship may have up to 18 decks. Swimming pools on the sundeck allow passengers to float aboard the floating vessel.

Rudder
A rudder controls the ship's direction.

Propeller
Twin propellers drive the ship.

Engine room
Located near the bottom of the ship towards its rear (aft), the engine room holds the machinery that drives the ship.

Stabilizer
Horizontal stabilizers prevent the ship from rolling side to side.

Hull
Welded construction maximizes the strength of the hull. Some ships are designed with a stronger double hull.

Relative density

Objects that are less dense than water float, while denser items sink. This is known as relative density. Pure water has a density of $1\,g/cm^3$. People, icebergs, and most types of wood float because their densities are less than $1\,g/cm^3$.

Cork has a very low density, so it floats high in the water.

A goldfish has a swim bladder full of air, which it uses to regulate its density, allowing it to float at different depths.

Nearly all metals are denser than water so sink like this steel bolt.

Floating city

Vast cruise ships can carry nearly 10,000 people, along with fuel, food, water, and cargo (known as dead weight), and the ship's machinery (lightweight), displacing 100,000 tonnes of water. How can these juggernauts of the sea float?

How boats float

Water exerts pressure on any object immersed in it. Pressure increases with depth, so the pressure on the underneath of an object is greater than the pressure on the top. The difference results in a force known as upthrust, or buoyancy. If the upthrust on a submerged object is equal to the object's weight, the object will float.

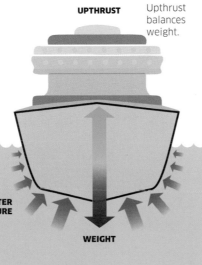

UPTHRUST

Upthrust balances weight.

UPTHRUST

WATER PRESSURE

WEIGHT

Sink or swim
A solid block of steel sinks because its weight is greater than upthrust, but a steel ship of the same weight floats because its hull is filled with air so its density overall is less than the density of water.

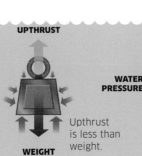

UPTHRUST

WEIGHT

Upthrust is less than weight.

Floating in air

Like water, air exerts pressure on objects with a force called upthrust that equals the weight of air pushed aside by the object. Few objects float in air because it is light, but the air in hot-air balloons is less dense than cool air.

Turbofan jet engine
A large fan sucks air into the engine. Some air is compressed before flowing into a combustion chamber. There it mixes with fuel and ignites to create hot exhaust gases that leave the engine at high velocity, pushing the plane forwards. Most air bypasses the engine at a lower velocity, but still contributes to thrust.

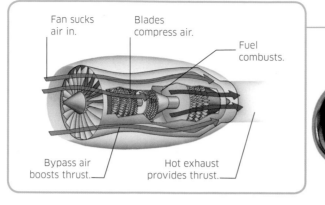

Fan sucks air in.

Blades compress air.

Fuel combusts.

Bypass air boosts thrust.

Hot exhaust provides thrust.

Ailerons
Left and right ailerons are moved up or down to raise or lower the wings; this is known as roll.

Radar

Pilot's seat

Front landing gear

Flight

Dynamics is the science of movement, and aerodynamics is movement through air. In order to fly, planes use thrust and lift to counteract the forces of drag and gravity.

Just over a hundred years since the first powered flight, today more than 100,000 planes fly every day and it seems normal to us that an airliner weighing as much as 562 tonnes when laden can take to the skies. To take off, a plane must generate enough lift to overcome gravity, using the power of its engines to create drag-defying thrust.

Airbus A380
The Airbus 380 is the world's biggest passenger airliner: 73 m (234 ft) long with a wing span of 79.8 m (262 ft), it can seat 555 people on two decks and carry 150 tonnes of cargo.

The forces of flight

Four forces act upon an aeroplane travelling through the air: thrust, lift, gravity, and drag. Thrust from the engines pushes the plane forwards, forcing air over the wings, which creates lift to get it off the ground, while gravity pulls the plane downwards, and drag – or air resistance – pulls it backwards. In level flight at a constant speed, all four of these forces are perfectly balanced.

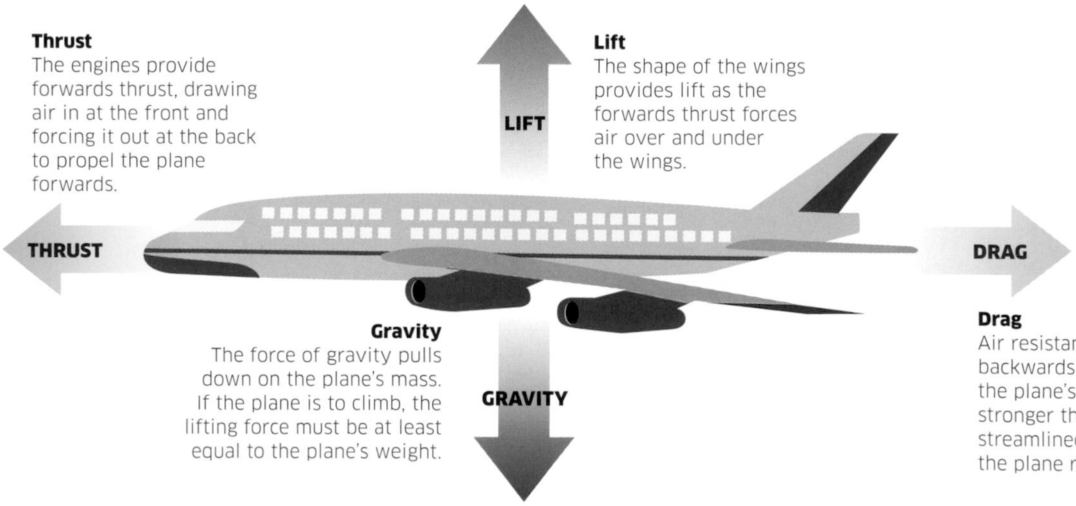

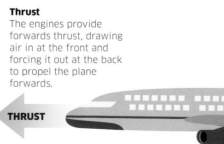

Thrust
The engines provide forwards thrust, drawing air in at the front and forcing it out at the back to propel the plane forwards.

Lift
The shape of the wings provides lift as the forwards thrust forces air over and under the wings.

LIFT

THRUST

DRAG

Gravity
The force of gravity pulls down on the plane's mass. If the plane is to climb, the lifting force must be at least equal to the plane's weight.

GRAVITY

Drag
Air resistance pulls backwards. The greater the plane's speed, the stronger the drag. The streamlined shape of the plane reduces drag.

262 tonnes - the **maximum fuel capacity** of an Airbus A380.

280 km/h (180 mph) - the **average take-off speed** of a jet airliner.

111

Aerofoil

The cross-section of a plane's wing has a shape called an aerofoil, which forces air to speed up over the top surface and slow down beneath. The aerofoil is angled so that air passing under the wing is forced downwards. Air passing over the wing is forced downwards too. The angle also creates an area of very low pressure above the wing. As a result of the wing pushing the air downwards and the pressure difference above and below the wing, the air pushes the wing (and plane) upwards.

Lower-pressure air above wing

Difference in pressure generates lift.

Higher-pressure air below wing

Gravity counteracts lift.

Angle of attack
The streamlined shape of the aerofoil is angled downwards towards the rear of the plane, allowing air to move smoothly over it. This is known as the angle of attack.

Air passing over and under the wing is forced downwards.

Vertical stabilizer

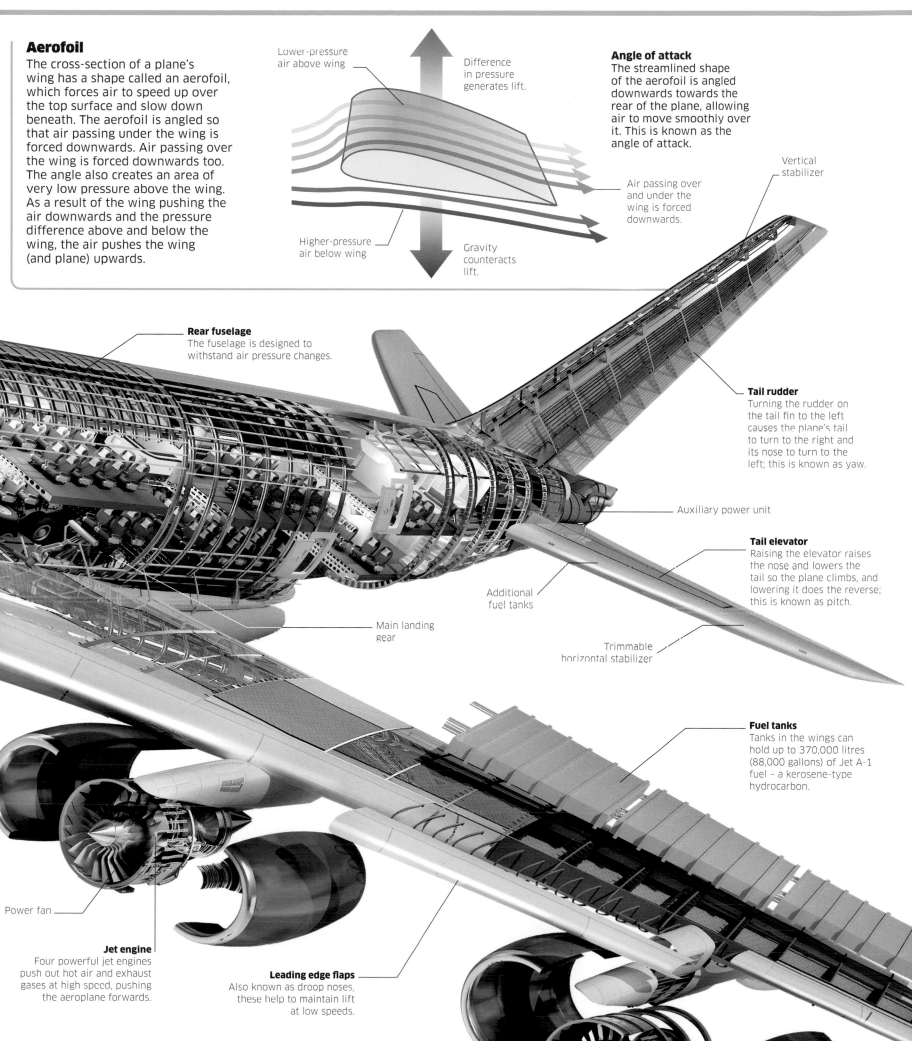

Rear fuselage
The fuselage is designed to withstand air pressure changes.

Tail rudder
Turning the rudder on the tail fin to the left causes the plane's tail to turn to the right and its nose to turn to the left; this is known as yaw.

Auxiliary power unit

Tail elevator
Raising the elevator raises the nose and lowers the tail so the plane climbs, and lowering it does the reverse; this is known as pitch.

Additional fuel tanks

Main landing gear

Trimmable horizontal stabilizer

Fuel tanks
Tanks in the wings can hold up to 370,000 litres (88,000 gallons) of Jet A-1 fuel - a kerosene-type hydrocarbon.

Power fan

Jet engine
Four powerful jet engines push out hot air and exhaust gases at high speed, pushing the aeroplane forwards.

Leading edge flaps
Also known as droop noses, these help to maintain lift at low speeds.

SPACE AND EARTH

All of space, matter, energy, and time make up the Universe – a vast, ever-expanding creation that is so big it would take billions of years to cross it, even when travelling at the speed of light. Within the Universe are clumps of matter called galaxies, and within those are planets like our own – Earth.

THE EXPANSION OF SPACE

Astronomers on Earth can observe galaxies moving away from us, but in reality they are moving away from every other point in the Universe as well. These galaxies are not moving into new space – all of space is expanding and pulling them away from each other. This effect can be imagined by thinking of the Universe as a balloon. As the balloon inflates, the rubber stretches and individual points on it all move further away from each other.

Huge expanses of space will come between galaxies in the future.

Although the galaxies remain the same size, the distance between them has grown,

Galaxies used to be more tightly clumped together.

BILLIONS OF YEARS AGO

TODAY

BILLIONS OF YEARS IN THE FUTURE

THE BIG BANG

The Universe came into existence around 13.8 billion years ago in a cataclysmic explosion known as the Big Bang. Starting out as tinier than an atom, it rapidly expanded – forming stars, and clusters of stars called galaxies. A large part of this expansion happened incredibly quickly – it grew by a trillion kilometres in under a second.

The Universe is dark until stars form.

Stars form.

379,000 years after the Big Bang, this afterglow light was emitted. It can still be seen in the Universe today.

Expansion quickly happens.

The Universe begins from nothing.

First galaxies form.

1 The Universe suddenly appears. At this stage, it is made up of pure energy and reaches extreme temperatures.

2 Rapid expansion (inflation) takes place, transforming the Universe from a tiny mass smaller than a fraction of an atom into a gigantic space the size of a city.

3 Matter is created from the Universe's energy. This starts out as minuscule particles and antiparticles (with the same mass as particles but with an opposite electric charge). Many of these converge and cancel each other out, but some matter remains.

4 The Universe is still less than a second old when the first recognizable subatomic particles start to form. These are protons and neutrons – the particles that make up the nucleus of an atom.

5 Over the next 379,000 years, the Universe slowly cools, until eventually atoms are able to form. This development changes the Universe from a dense fog into an empty space punctuated by clouds of hydrogen and helium gas. Light can now pass through it.

THE OBSERVABLE UNIVERSE

When we look at distant objects in the night sky, we are actually seeing what they looked like millions, or even billions, of years ago, as that is how long the light from them has taken to reach us. All of the space we can see from Earth is known as the observable Universe. Other parts lie beyond that, but are too far away for the light from them to have reached us yet. However, using a space-based observatory such as the Hubble Space Telescope, we can capture images of deep space and use them to decipher the Universe's past.

Hubble imaging
The Hubble Space Telescope has been operating since 1990 and has captured thousands of images of the Universe. Many of these have been compiled to create amazing views of the furthest (and therefore oldest) parts of the Universe we can see. These are known as Deep Field images.

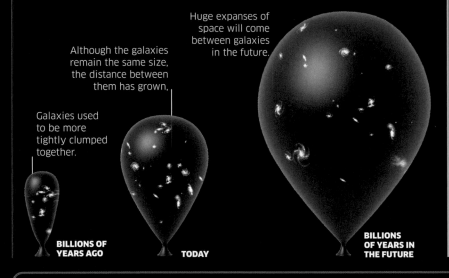

HUBBLE DEEP FIELD

HUBBLE ULTRA DEEP FIELD

FIRST GALAXIES

DARK AGES

RADIATION ERA

The first Hubble Deep Field observed one part of the night sky over 10 days. It revealed galaxies formed less than a billion years after the Big Bang.

The later Hubble Ultra Deep Field image (above) shows even further into the past, picturing galaxies formed 13 billion years ago, when the Universe was around 800 million years old.

There are regions of space further back in time that Hubble and other powerful space telescopes cannot see.

6 Just over half a million years after the Big Bang, the distribution of matter in the Universe begins to change. Tiny denser patches of matter begin to be pulled closer together by gravity.

8 Stars form in groups within vast clouds of gas. The first groups become the first galaxies. Most of these are relatively small, but later merge to form larger galaxies that stretch for hundreds of millions of light years.

10 Our Solar System comes into being after 9 billion years, formed from the collapse of a large nebula (a cloud of gas and dust). Material first forms into the Sun, and then other clumps become the variety of planets surrounding it, including Earth.

A NASA probe was launched in 2001 to measure the size and properties of the Universe.

11 In the future, the Universe will continue to expand and change, and our Solar System will not last for ever. The Sun is very slowly getting hotter, and when the Universe is 20 billion years old, it will also expand in size – an event likely to destroy Earth.

7 The effects of gravity begin to create more and more clumps of matter, until large spheres of gas, called stars, are formed. The Universe is now 300 million years old. These stars produce the energy to sustain themselves by nuclear fusion.

9 Around 8 billion years after the Big Bang, the expansion of the Universe begins to accelerate.

Solar System forms.

12 Scientists do not know exactly how the Universe will end, but it is predicted to keep on expanding and become incredibly cold and dark in a process known as the "Big Chill".

Redshift
When an object (a distant galaxy) is moving away from the observer (us), its wavelengths get longer. The light it produces therefore shifts into the red end of the light spectrum. More distant galaxies have greater redshift, supporting the theory that the Universe is expanding.

Blueshift
A few nearby galaxies are actually moving towards us. Their wavelengths will be shorter, shifting the light they produce to the blue end of the spectrum.

DISCOVERING THE BIG BANG

Scientists did not always believe in the theory of an expanding Universe and the Big Bang. However, during the 20th century, several discoveries were made which supported this idea. In 1929, American astronomer Edwin Hubble observed that the light coming from distant galaxies appeared redder than it should be. He attributed this to a phenomenon called redshift, suggesting that galaxies must be moving away from us. Another piece of evidence was the discovery of Cosmic Background Radiation – microwaves coming from all directions in space that could only be explained as an after effect of the Big Bang.

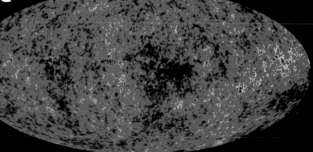

Cosmic Background Radiation
This image, captured by NASA's Wilkinson Microwave Anisotropy Probe, shows a false colour depiction of the background radiation that fills the entire Universe. This is the remains of the intense burst of energy that was released by the Big Bang.

Most of a **galaxy's mass** is made up of **dark matter**.

It is estimated that there are **2 trillion galaxies** in the parts of **the Universe** we can see.

Galaxies

Unimaginably huge collections of gas, dust, stars, and even planets, galaxies come in many shapes and sizes. Some are spirals, such as our own galaxy, others are like squashed balls, and some have no shape at all.

When you look up at the sky at night, every star you see is part of our galaxy, the Milky Way. This is part of what we call the Local Group, which contains about 50 galaxies. Beyond it are countless more galaxies that stretch out as far as telescopes can see. The smallest galaxies in the Universe have a few million stars in them, while the largest have trillions. The Milky Way lies somewhere in the middle, with between 100 billion and a trillion stars in it. The force of gravity holds the stars in a galaxy together, and they travel slowly around the centre. A supermassive black hole hides at the heart of most galaxies.

Astronomers have identified four types of galaxies: spiral, barred spiral, elliptical, and irregular. Spiral galaxies are flat spinning disks with a bulge in the centre, while barred spiral galaxies have a longer, thinner line of stars at their centre, which looks like a bar. Elliptical galaxies are an ellipsoid, or the shape of a squashed sphere – these are the largest galaxies. Then there are irregular galaxies, which have no regular shape.

MILKY WAY

Type: *Barred spiral*
Diameter: *100,000 light years*

Our own galaxy, the Milky Way, is thought to be a barred spiral shape, but we cannot see its shape clearly from Earth because we are part of it. From our Solar System, it appears as a pale streak in the sky with a central bulge of stars. From above, it would look like a giant whirlpool that takes 200 million years to rotate.

Galactic core
Infrared and X-ray images reveal intense activity near the galactic core. The galaxy's centre is located within the bright white region. Hundreds of thousands of stars that cannot be seen in visible light swirl around it, heating dramatic clouds of gas and dust.

Solar system
Our solar system is in a minor spiral arm called the Orion arm.

Side view of the Milky Way
Viewed from the side, the Milky Way would look like two fried eggs back to back. The stars in the galaxy are held together by gravity and travel slowly around the galactic heart in a flat orbit.

Our **Sun** lies between **25,000** and **28,000 light years** from the **centre of the Milky Way**.

The **largest galaxies** in the Universe stretch up to **2 million light years long**.

The word **galaxy** comes from the **Greek** term *galaxias kyklos*, which means **milky circle**.

115

ANDROMEDA

Type: Spiral
Distance: 2,450,000 light years

Our closest large galaxy, Andromeda – a central hub surrounded by a flat, rotating disc of stars, gas, and dust – can sometimes be seen from Earth with the naked eye. In 4.5 billion years, Andromeda is expected to collide with the Milky Way, forming one huge elliptical galaxy.

MESSIER 87

Type: Elliptical
Distance: 53 million light years

M87, also known as Virgo A, is one of the largest galaxies in our part of the Universe. The galaxy is giving out a powerful jet of material from the supermassive black hole at its centre, energetic enough to accelerate particles to nearly the speed of light.

SMALL MAGELLANIC CLOUD

Type: Irregular dwarf
Distance: 197,000 light years

The dwarf galaxy SMC stretches 7,000 light years across. Like its neighbour the Large Magellanic Cloud (LMC), its shape has been distorted by the gravity of our own galaxy. Third closest to the Milky Way, it is known as a satellite galaxy because it orbits our own.

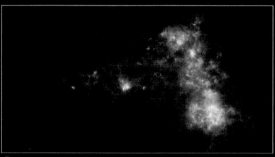

CARTWHEEL GALAXY

Type: Irregular ring
Distance: 500 million light years

The Cartwheel Galaxy started out as a spiral. However, 200 million years ago it collided with a smaller galaxy, causing a powerful shock throughout the galaxy, which tossed lots of the gas and dust to the outside, creating its unusual shape.

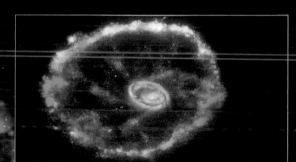

ANTENNAE GALAXIES

Type: Merging spirals
Distance: 45-65 million light years

Around 1.2 billion years ago, the Antennae Galaxies were two separate galaxies: one barred spiral and one spiral. They started to merge a few hundred million years ago, when the antennae formed and are expected to become one galaxy in about 400 million years.

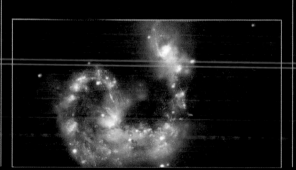

WHIRLPOOL GALAXY

Type: Colliding spiral and dwarf
Distance: 23 million light years

About 300 million years ago, the spiral Whirlpool Galaxy was struck by a dwarf galaxy, which now appears to dangle from one of its spiral arms. The collision stirred up gas clouds, triggering a burst of star formation, which can be seen from Earth with a small telescope.

Active galaxies
Some galaxies send out bright jets of light and particles from their centres. These "active" galaxies can be grouped into four types: radio galaxies, Seyfert galaxies, quasars, and blazars. All are thought to have supermassive black holes at their core, known as the active galactic nuclei, which churn out the jets of material.

Two strong jets spurt out of the supermassive black hole.

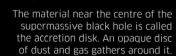

The material near the centre of the supermassive black hole is called the accretion disk. An opaque disc of dust and gas gathers around it.

The jets have so much energy they move at nearly the speed of light.

Our Sun is a star – it only **seems bigger** than other stars in the night sky because it is much **closer to Earth.**

1 Interstellar cloud
Stars are born in huge clouds of dense, cold gas and dust. A supernova explosion or star collision can trigger star birth.

2 Fragments form
The cloud breaks up into fragments. Gravity pulls the most massive and dense of these into clumps.

3 Protostar forms
Gravity pulls more material into the protostar's core. Density, pressure, and temperature build up.

4 Spinning disc
The material being pulled in starts to spin round, blowing out jets of gas.

Star life cycle

Stars are born in vast clouds of cold, dense interstellar gas and dust that evolve until, billions of years later, they run out of fuel and die.

The clouds that give birth to stars consist mainly of hydrogen gas. New stars are huge, spinning globes of hot, glowing gas – mainly hydrogen with some helium. Most of this material is packed into the stars' cores, setting off nuclear reactions, fuelled by hydrogen, that form helium and release energy in the form of heat and light. When most of the hydrogen is used up, stars may fade away, expand, or collapse in on themselves.

5 Main sequence star
The core becomes so hot and dense that nuclear reactions occur and the star shines.

6 Planets form
Debris spinning around the star may clump together to form planets, moons, comets, and asteroids.

7 Stable star
The glowing core produces an outward pressure that balances the inward pull of gravity.

Birth, life, and death of a star

Stars start out their life as clouds of gas and dust, called nebulae. After millions of years, these clouds begin to pull inwards because of the gravity of the gas and dust. As it is squeezed, the cloud heats up to form a young star, known as a protostar. If this reaches 15 million degrees Celsius, it is hot enough to start nuclear fusion – the reaction needed for a star to form. The energy produced prevents a star from collapsing under its own weight and makes it shine. What happens when the fuel runs out and the star dies depends on how much dust gathered in the first place.

The Sun has existed for about
4.5 billion years,
and has burnt about half of
its hydrogen fuel.

Death of a small star
Stars with less than half the mass of the Sun, called red dwarfs, fade away slowly. Once the hydrogen in the core is used up, the star begins to feed off hydrogen in its atmosphere, shrinking – over up to a trillion years – to become a black dwarf.

Black dwarf
When all fuel is used up and its light is extinguished, the star becomes a cinder the size of Earth.

Star continues to shrink and fade.

Light intensity fades out.

Star begins to shrink.

Death of a medium-sized star
When a star with the same mass as our Sun has used up its hydrogen, after about 10 billion years, nuclear fusion spreads out from the core, making the star expand into a red giant. The core collapses until it is hot and dense enough to fuse helium. When this too runs out, the star becomes a white dwarf, its outer layers spreading into space as a cloud of debris.

Neutron stars are the smallest, most dense stars in the Universe - 10 km (6 miles) in diameter but with up to 30 times as much mass as our Sun.

Energy released in the centre of the Sun takes millions of years to reach its surface.

117

If you sorted all the stars into piles, the biggest pile, by far, would be

red dwarfs – stars

with less than half of the Sun's mass.

Death of a massive star

Stars more than eight times the mass of our Sun will be hot enough to become supergiants. The heat and pressure in the core become so intense that nuclear fusion can fuse helium and larger atoms to create elements such as carbon or oxygen. As this happens, the stars swell into supergiants, which end their lives in dramatic explosions called supernovae. Smaller supergiants become neutron stars, but larger ones become black holes.

Red supergiant
Nuclear fusion carries on inside the core of the supergiant, forming heavy elements until the core turns into iron and the star collapses.

Star types

The Hertzsprung–Russell diagram is a graph that astronomers use to classify stars. It plots the brightness of stars against their temperature to reveal distinct groups of stars, such as red giants (dying stars) and main sequence stars (ordinary stars). Astronomers also classify stars by colour, which relates to temperature. Red is the coolest colour, seen in stars cooler than 3,500°C (6,000°F). Stars such as our Sun are yellowish white and average around 6,000°C (10,000°F). The hottest stars are blue, with surface temperatures above 12,000°C (21,000°F).

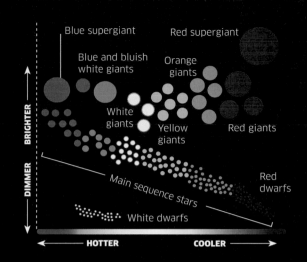

Blue supergiant — Red supergiant
Blue and bluish white giants — Orange giants
White giants — Yellow giants — Red giants
Red dwarfs
Main sequence stars
White dwarfs

BRIGHTER — DIMMER

◄— HOTTER — COOLER —►

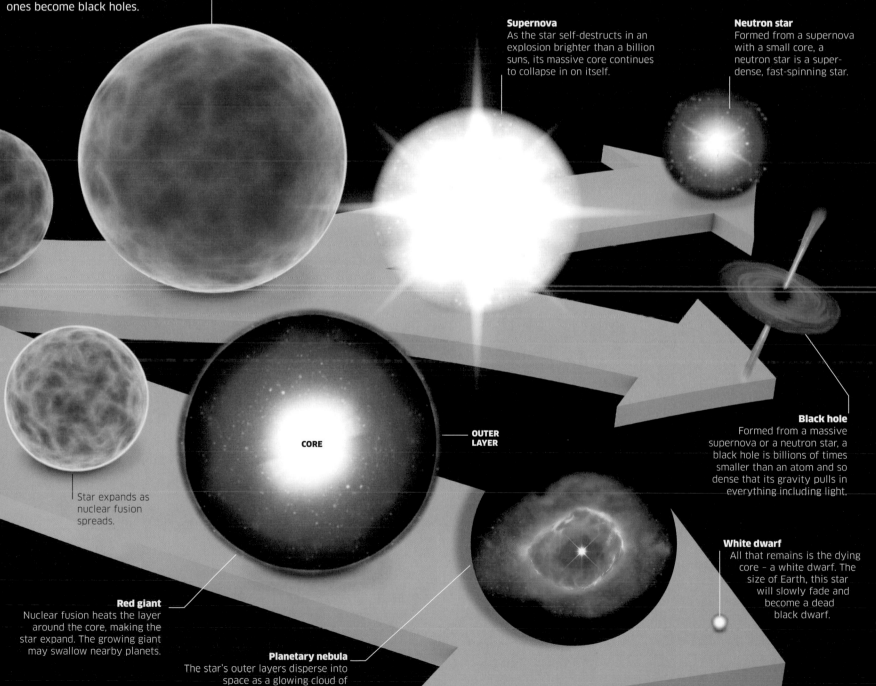

Supernova
As the star self-destructs in an explosion brighter than a billion suns, its massive core continues to collapse in on itself.

Neutron star
Formed from a supernova with a small core, a neutron star is a super-dense, fast-spinning star.

CORE

OUTER LAYER

Black hole
Formed from a massive supernova or a neutron star, a black hole is billions of times smaller than an atom and so dense that its gravity pulls in everything including light.

Star expands as nuclear fusion spreads.

White dwarf
All that remains is the dying core – a white dwarf. The size of Earth, this star will slowly fade and become a dead black dwarf.

Red giant
Nuclear fusion heats the layer around the core, making the star expand. The growing giant may swallow nearby planets.

Planetary nebula
The star's outer layers disperse into space as a glowing cloud of wreckage – a planetary nebula. The material in this cloud will eventually be recycled to form new stars.

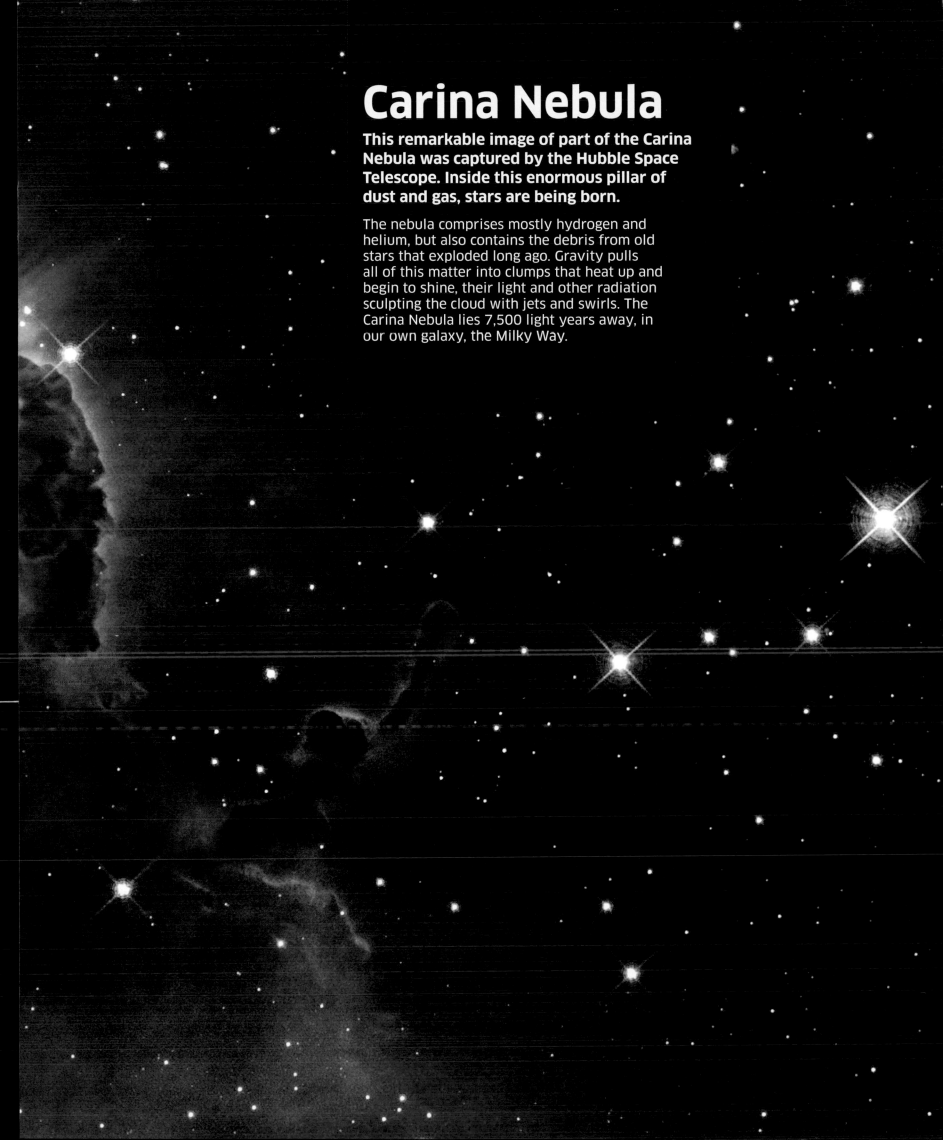

Carina Nebula

This remarkable image of part of the Carina Nebula was captured by the Hubble Space Telescope. Inside this enormous pillar of dust and gas, stars are being born.

The nebula comprises mostly hydrogen and helium, but also contains the debris from old stars that exploded long ago. Gravity pulls all of this matter into clumps that heat up and begin to shine, their light and other radiation sculpting the cloud with jets and swirls. The Carina Nebula lies 7,500 light years away, in our own galaxy, the Milky Way.

Size comparison
With a diameter of nearly 1.4 million km (870,000 miles), the Sun is 10 times wider than Jupiter, the biggest of the planets, and over 1,000 times more massive.

Inner planets
The inner four planets are smaller than the outer four. They are called the rocky planets.

MERCURY VENUS EARTH MARS

Outer planets
The outermost four planets are larger and made up of gas, so they are called the gas giants.

JUPITER   SATURN

URANUS   NEPTUNE

Oort Cloud
The Oort Cloud is a ring of tiny, icy bodies that is thought to extend between 50,000 and 100,000 times further from the Sun than the distance from the Sun to Earth – but it's so far away that no-one really knows.

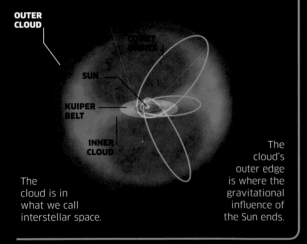

OUTER CLOUD

COMET ORBITS

SUN

KUIPER BELT

INNER CLOUD

The cloud is in what we call interstellar space.

The cloud's outer edge is where the gravitational influence of the Sun ends.

Distance from the Sun
It is hard to imagine how far Earth is from the Sun, and how much bigger the Sun is than Earth. If Earth was a peppercorn, the Sun would be the size of a bowling ball – 100 times bigger.

Kuiper Belt
The Solar System does not end beyond Neptune: the Kuiper Belt (30–55 AU from the Sun) is home to smaller bodies that include dwarf planets.

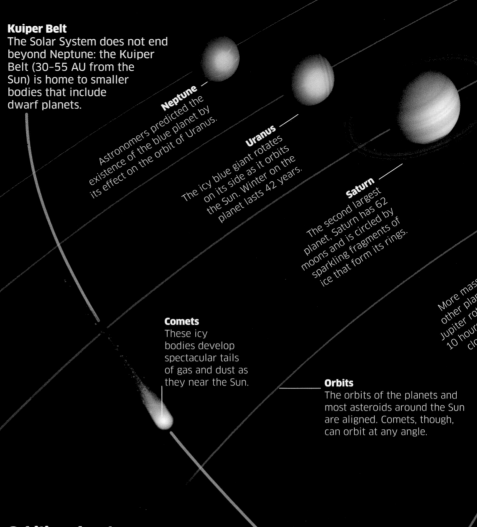

Neptune
Astronomers predicted the existence of the blue planet by its effect on the orbit of Uranus.

Uranus
The icy blue giant rotates on its side as it orbits the Sun. Winter on the planet lasts 42 years.

Saturn
The second largest planet, Saturn has 62 moons and is circled by sparkling fragments of ice that form its rings.

Jupiter
More massive than the other planets combined, Jupiter rotates once every 10 hours, whipping its red clouds into stripes and swirling storms.

Comets
These icy bodies develop spectacular tails of gas and dust as they near the Sun.

Orbits
The orbits of the planets and most asteroids around the Sun are aligned. Comets, though, can orbit at any angle.

Orbiting planets
There are eight planets in the solar system. They form two distinct groups. The inner planets – Mercury, Venus, Earth, and Mars – are solid balls of rock and metal. The outer planets – Jupiter, Saturn, Uranus, and Neptune – are gas giants: enormous, swirling globes made mostly of hydrogen and helium.

The Solar System
The Solar System is a huge disc of material, with the Sun at its centre, that stretches out over 30 billion km (19 billion miles) to where interstellar space begins.

Most of the Solar System is empty space, but scattered throughout are countless solid objects bound to the Sun by gravity and orbiting around it. These include the eight planets, hundreds of moons and dwarf planets, millions of asteroids, and possibly billions of comets. The Sun itself makes up 99.8 per cent of the mass of the Solar System.

SUN MERCURY VENUS EARTH MARS   JUPITER   SATURN

Earth is 149.6 million km (92.9 million miles) from the Sun – or one astronomical unit (AU).

Jupiter is 780 million km (484 million miles) from the Sun, which is equal to 5.2 AU.

Saturn orbits on average 1.43 billion km (890 million miles) from the Sun, or 9.58 AU.

There are **five known dwarf planets :**
Ceres, Pluto, Makemake, Eris, and Haumea.

Asteroid 234 Ida
In between the orbits of Mars
and Jupiter lies the asteroid belt.
Asteroids are made up of a mixture
of rock and ice. This space rubble
is the detritus of planet
formation.

Sun
The Sun lies in the centre
of the Solar System. It spins
on its axis, taking less than
25 days to rotate despite
its massive size.

Venus
Venus rotates in the
opposite direction to the
other planets, so slowly
that it takes 224 days to
complete one rotation.

Mercury
The closest planet to the Sun,
Mercury is also the smallest.
It takes 88 days to make a trip
around the Sun, rotating three
times for every two orbits.

Earth
Our home planet, Earth is the
only planet we know of that
can support life, thanks to its
oceans and atmosphere.

Mars
Mars is a rocky planet, but it does
not have a magnetic field like
Earth's to deflect space radiation.

Orbit speed
The further a
planet is from
the Sun, the
slower it travels and
the longer its orbit takes.
The most distant planet,
Neptune, takes 165 years to
travel around the Sun, at
5.43 km/s (3.37 miles per second).

URANUS

NEPTUNE

Uranus is 2.87 billion km (1.78 billion miles)
from the Sun on average, or 19.14 AU.

Neptune orbits at 4.53 billion km (2.81 billion miles), an average
of 30 times the distance between Earth and the Sun, or 30 AU.

The seasons

As Earth orbits around the Sun, it also rotates around its axis – an imaginary north-south line. This axis is tilted by 23.4° compared to Earth's orbit, so that one part of the planet is always closer to or further away from the Sun, resulting in the seasons.

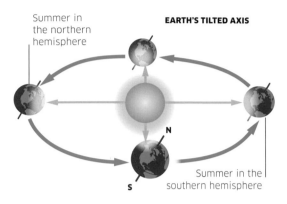

Summer in the northern hemisphere

EARTH'S TILTED AXIS

N

S

Summer in the southern hemisphere

Atmosphere

Earth's atmosphere is made up of a mix of gases – 78 per cent nitrogen, 21 per cent oxygen, and a small amount of others, such as carbon dioxide and argon. These gases trap heat on the planet and let us breathe. The atmosphere has five distinct layers.

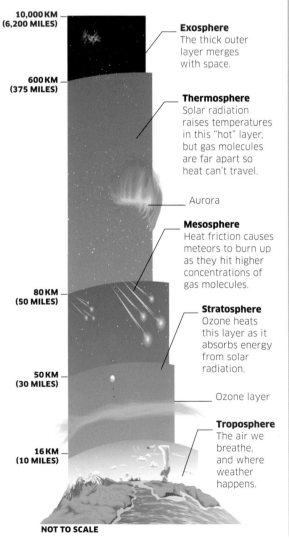

10,000 KM (6,200 MILES)

600 KM (375 MILES)

80 KM (50 MILES)

50 KM (30 MILES)

16 KM (10 MILES)

Exosphere
The thick outer layer merges with space.

Thermosphere
Solar radiation raises temperatures in this "hot" layer, but gas molecules are far apart so heat can't travel.

Aurora

Mesosphere
Heat friction causes meteors to burn up as they hit higher concentrations of gas molecules.

Stratosphere
Ozone heats this layer as it absorbs energy from solar radiation.

Ozone layer

Troposphere
The air we breathe, and where weather happens.

NOT TO SCALE

Earth and Moon

Our home, Earth, is about 4.5 billion years old. With a diameter of just over 12,000 km (7,500 miles), it orbits the Sun every 365.3 days and spins on its axis once every 23.9 hours.

Of all the planets in the Universe, ours is the only place life is known to exist. Earth is one of the Solar System's four rocky planets, and the third from the Sun. Its atmosphere, surface water, and magnetic field – which protects us from solar radiation – make Earth the perfect place to live.

Inside Earth

Earth is made up of rocky layers. The outer crust floats on a rocky shell called the mantle. Beneath this is the hot, liquid outer core and the solid inner core.

Outer core
The liquid outer layer of the Earth's core is hot. Made of liquid iron and nickel, it is 2,300 km (1,400 miles) thick.

Oceanic crust
The solid outer layer of rocks is the crust. Under the oceans, it is only about 10 km (6 miles) thick, but it is denser than the continental crust.

Continental crust
The continental crust is the land on which we stand. It is much thicker than the oceanic crust. It is up to 70 km (45 miles) thick, but less dense.

Sun
The Sun's diameter is 109 times Earth's.

Every year, the **Moon drifts 3.78 cm (1.48 in) further away from Earth**.

Earth's **inner core spins at a different speed** to the rest of the planet.

More than **300,000 impact craters** wider than **1 km (0.6 miles) cover the Moon's surface**.

123

The Moon
Orbiting Earth every 27 days, the Moon is a familiar sight in the night sky. The same side of the Moon always faces the Earth. The dark side of the Moon can only be seen from spacecraft.

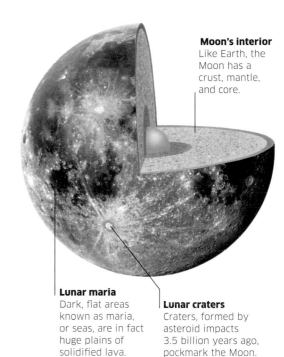

Inner core
The iron inner core is just over two-thirds of the size of the Moon and as hot as the surface of the Sun. It is solid because of the immense pressure on it.

Lower mantle
The lower layer of the mantle contains more than half the planet's volume and extends 2,900 km (1,800 miles) below the surface. It is hot and dense.

Upper mantle
The layer extending 410 km (255 miles) below the crust is mostly solid rock, but it moves as hot, molten rock rises to the surface, cools, and then sinks.

Earth
Earth's diameter is four times that of the Moon, and our planet weighs 80 times more than its satellite.

Earth to Sun
The Sun is 150 million km (93 million miles) from Earth. It takes light 8 minutes to travel this distance, known as one astronomical unit (AU).

Moon
The Moon is 384,000 km (239,000 miles) from Earth.

Moon
Our only natural satellite, the Moon is almost as old as Earth. It is thought it was made when a flying object the size of Mars crashed into our planet, knocking lots of rock into Earth's orbit. This rock eventually clumped together to form our Moon. It is the Moon's gravitational pull that is responsible for tides.

Moon's interior
Like Earth, the Moon has a crust, mantle, and core.

Lunar maria
Dark, flat areas known as maria, or seas, are in fact huge plains of solidified lava.

Lunar craters
Craters, formed by asteroid impacts 3.5 billion years ago, pockmark the Moon.

Lunar cycle
The Moon doesn't produce its own light. The Sun illuminates exactly half of the Moon, and the amount of the illuminated side we see depends upon where the Moon is in its orbit ar. This gives rise to the phenomenon known as the phases of the Moon.

NEW MOON (0 DAYS)

WAXING CRESCENT

FIRST QUARTER (DAY 7)

WAXING GIBBOUS

FULL MOON (DAY 14)

WANING GIBBOUS

LAST QUARTER (DAY 21)

WANING CRESCENT

NEW MOON (DAY 28)

124 energy and forces ○ **TECTONIC EARTH**

55 km (34 miles) – **the average width of the 3,000-km (1,850-miles) long East African Rift System** of active faults.

Tectonic Earth

Earth's surface is a layer of solid rock split into huge slabs called tectonic plates, which slowly shift, altering landscapes and causing earthquakes and volcanoes.

The tectonic plates are made up of Earth's brittle crust fused to the top layer of the underlying mantle, forming a shell-like elastic structure called the lithosphere. Plate movement is driven by convection currents in the lower, viscous layers of the mantle – known as the asthenosphere – when hot, molten rock rises to the surface and cooler, more solid rock sinks. Most tectonic activity happens near the edges of plates, as they move apart from, towards, or past each other.

Plates move at between 7 mm (¼ in) per year, one-fifth **the rate human fingernails grow,** and 150 mm (6 in) per year – the rate human hair grows.

Continental drift

Over millions of years, continents carried by different plates have collided to make mountains, combined to form supercontinents, or split up in a process called rifting. South America's east coast and Africa's west coast fit like pieces of a jigsaw puzzle. Similar rock and life forms suggest that the two continents were once a supercontinent.

PANGAEA

270 MILLION YEARS AGO

LAURASIA

GONDWANA

180 MILLION YEARS AGO

NORTH AMERICA

AFRICA

SOUTH AMERICA

66 MILLION YEARS AGO

Plate tectonics

Where plates meet, landscape-changing events, such as island formation, rifting (separation), mountain-building, volcanic activity, and earthquakes take place. Plate boundaries fall into three main classes: divergent, convergent, and transform.

Divergent boundary
As two plates move apart, magma welling up from the mantle fills the gap and creates new plate. Linked with volcanic activity, divergent boundaries form mid-ocean spreading ridges under the sea.

Island arc
A series of underwater volcanoes forms a chain of islands, or an archipelago.

Ocean trench
Two ocean plates subduct to form a deep-sea trench.

Mid-ocean ridge
Magma wells up as plates move apart, forming a ridge on the ocean floor.

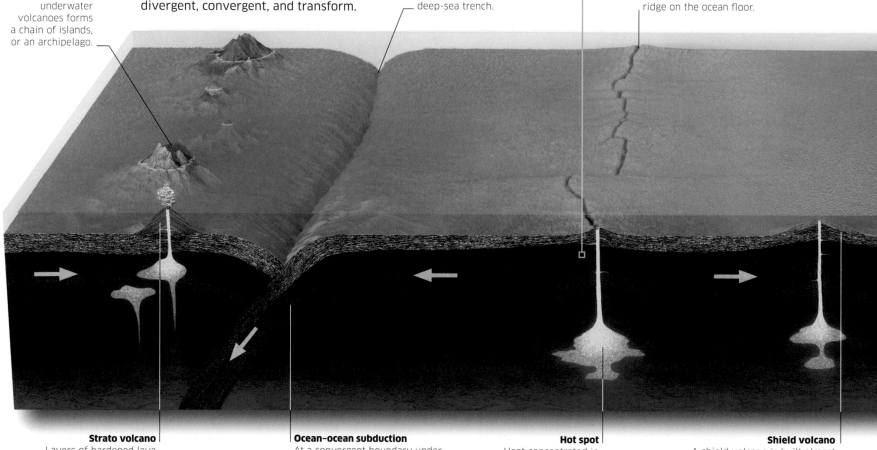

Strato volcano
Layers of hardened lava and ash build up, making these volcanoes steeper than shield volcanoes.

Ocean–ocean subduction
At a convergent boundary under the sea, one oceanic plate slides under the other, creating a mid-ocean trench.

Hot spot
Heat concentrated in some areas of the mantle can erupt as molten magma.

Shield volcano
A shield volcano is built almost entirely of very fluid lava flows, making it quite shallow in shape.

4 billion years old – the age of the oldest parts of Earth's crust.

9.5 The magnitude of the **largest ever recorded earthquake** in the world, **in Chile on 22 May 1960.**

452 The number of **volcanoes** around the **Pacific Ring of Fire.**

125

Tectonic plates

There are seven large plates and numerous medium-sized and smaller plates, which roughly coincide with the continents and oceans. The Ring of Fire is a zone of earthquakes and volcanoes around the Pacific plate from California in the northeast to Japan and New Zealand in the southwest.

━━━━━ **CONVERGENT**
━━━━━ **DIVERGENT**
───── **TRANSFORM**
╌╌╌╌╌ **UNCERTAIN**

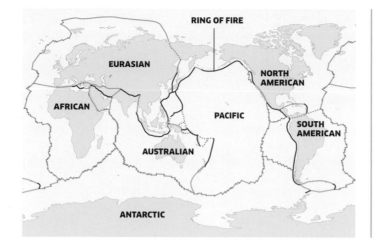

Colliding continents

When continents collide, layers of rock are pushed up into mountain ranges. Continental convergence between Indian subcontinent and Eurasian landmass formed the Himalayas.

Convergent boundary
As two plates move towards each other, one plate moves down, or subducts, under the other and is destroyed. A deep-sea trench or chain of volcanoes may form, and earthquakes often occur.

Transform boundary
When plate edges scrape past each other, earthquakes are frequent. The San Andreas fault in California is one famous example.

Sliding plates
Plates sliding past each other may make earthquakes happen.

Volcanic ranges
A chain of volcanoes develops on the side of the plate that is not subducting.

Rift valley
A valley appears where two plates move apart, or rift.

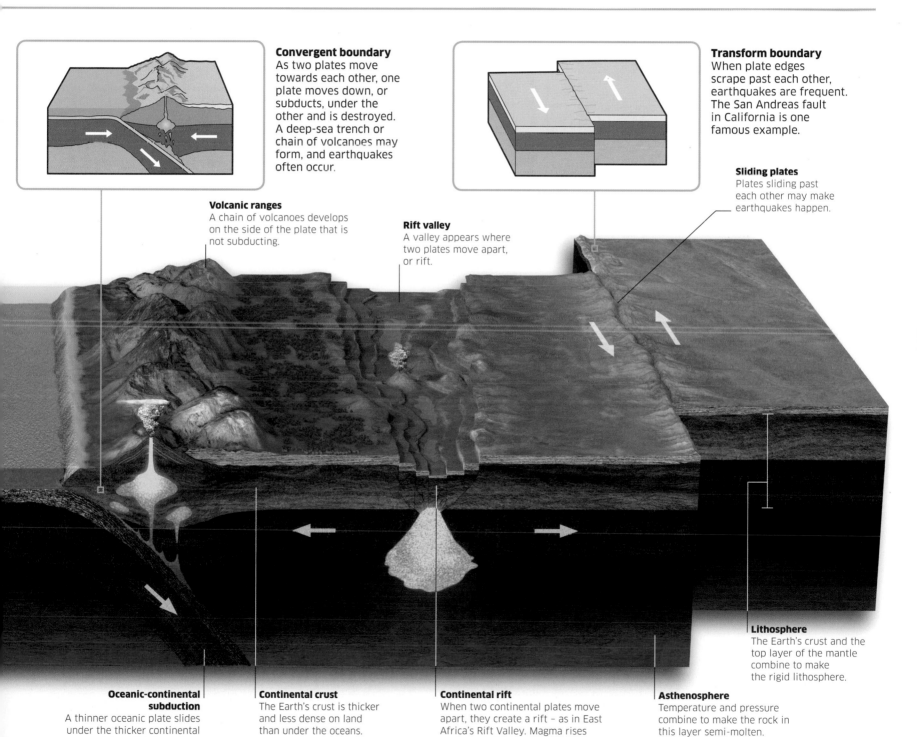

Oceanic-continental subduction
A thinner oceanic plate slides under the thicker continental plate at this boundary.

Continental crust
The Earth's crust is thicker and less dense on land than under the oceans.

Continental rift
When two continental plates move apart, they create a rift – as in East Africa's Rift Valley. Magma rises up through the gap, leading to volcanic activity.

Asthenosphere
Temperature and pressure combine to make the rock in this layer semi-molten.

Lithosphere
The Earth's crust and the top layer of the mantle combine to make the rigid lithosphere.

Supercell storms can **last 12 hours** and **travel 800 km (500 miles)**.

Storm clouds

Dense, dark clouds gathering overhead mean stormy weather is on the way. Thunderstorms have terrifying power but also an awesome beauty.

Our weather is created by changes in the atmosphere. When air turns cold, it sinks, becoming compressed under its own weight and causing high pressure at Earth's surface. As the air molecules squeeze together, they heat up. The warm air rises, surface pressure drops, and fair weather may follow. But when rapidly rising warm air meets descending cold air, the atmosphere becomes unsettled. Water vapour in the air turns into clouds, the clouds collide, and electric energy builds up. The electricity is released in lightning bolts that strike Earth's surface with cracks of thunder, often accompanied by heavy rainfall.

Supercell storms

One of the most dangerous weather conditions is the supercell storm, when a huge mass of cloud develops a rotating updraft of air, called a mesocyclone, at its centre. The cloud cover may stretch from horizon to horizon. Above this, unseen from the ground, a cloud formation known as cumulonimbus towers like a monstrous, flat-topped mushroom into the upper atmosphere. A supercell storm system can rage for many hours, producing destructive winds, torrential rain, and giant hailstones.

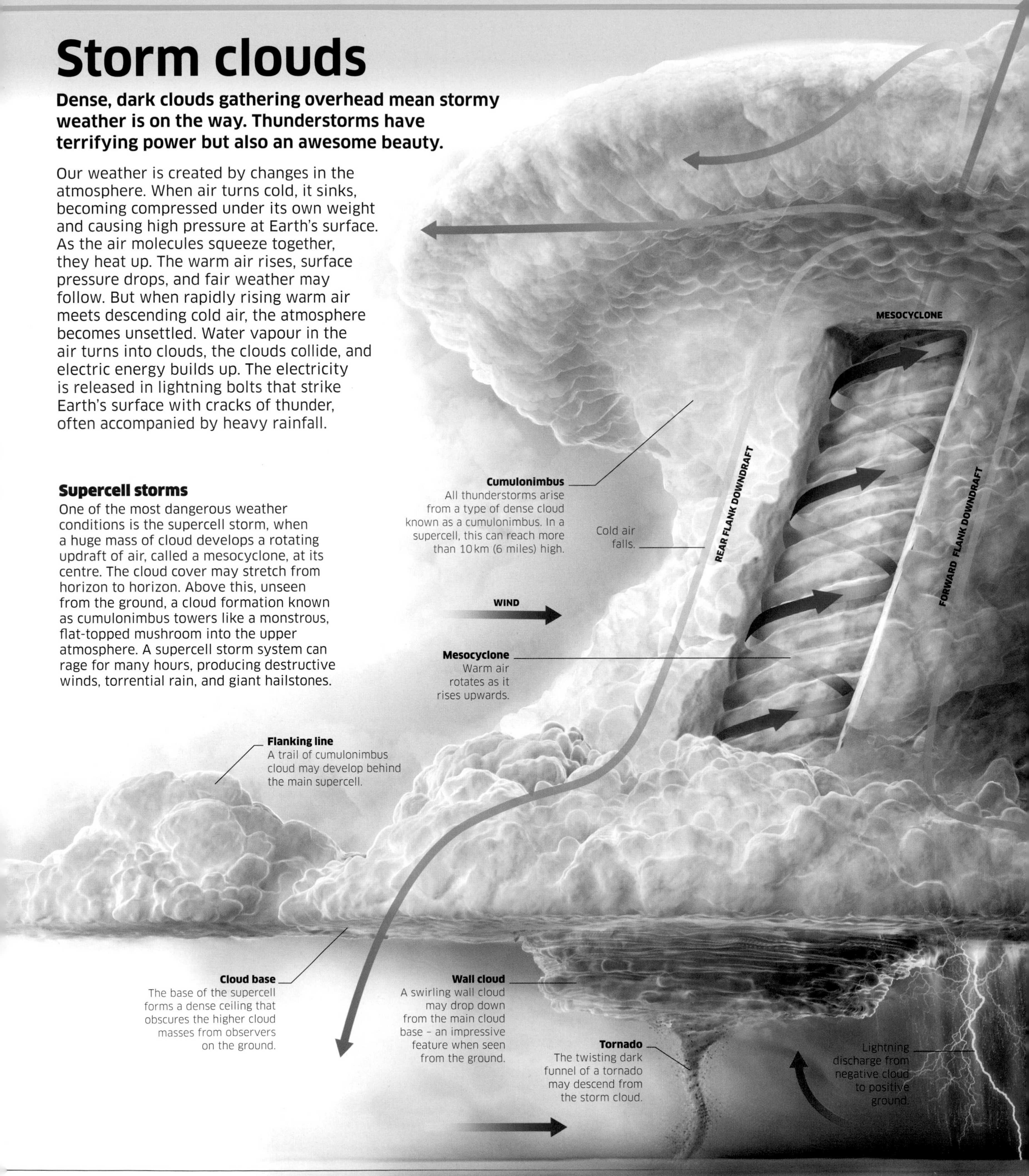

MESOCYCLONE

Cumulonimbus
All thunderstorms arise from a type of dense cloud known as a cumulonimbus. In a supercell, this can reach more than 10 km (6 miles) high.

Cold air falls.

REAR FLANK DOWNDRAFT

FORWARD FLANK DOWNDRAFT

WIND

Mesocyclone
Warm air rotates as it rises upwards.

Flanking line
A trail of cumulonimbus cloud may develop behind the main supercell.

Cloud base
The base of the supercell forms a dense ceiling that obscures the higher cloud masses from observers on the ground.

Wall cloud
A swirling wall cloud may drop down from the main cloud base – an impressive feature when seen from the ground.

Tornado
The twisting dark funnel of a tornado may descend from the storm cloud.

Lightning discharge from negative cloud to positive ground.

1 billion volts of electricity can be discharged by a **lightning bolt**.

At a temperature of 29,730°C (53,540°F), lightning is hotter than the surface of the Sun.

Lightning **"bolts from the blue"** can strike up to 25 km (15 miles) from a thunderstorm.

127

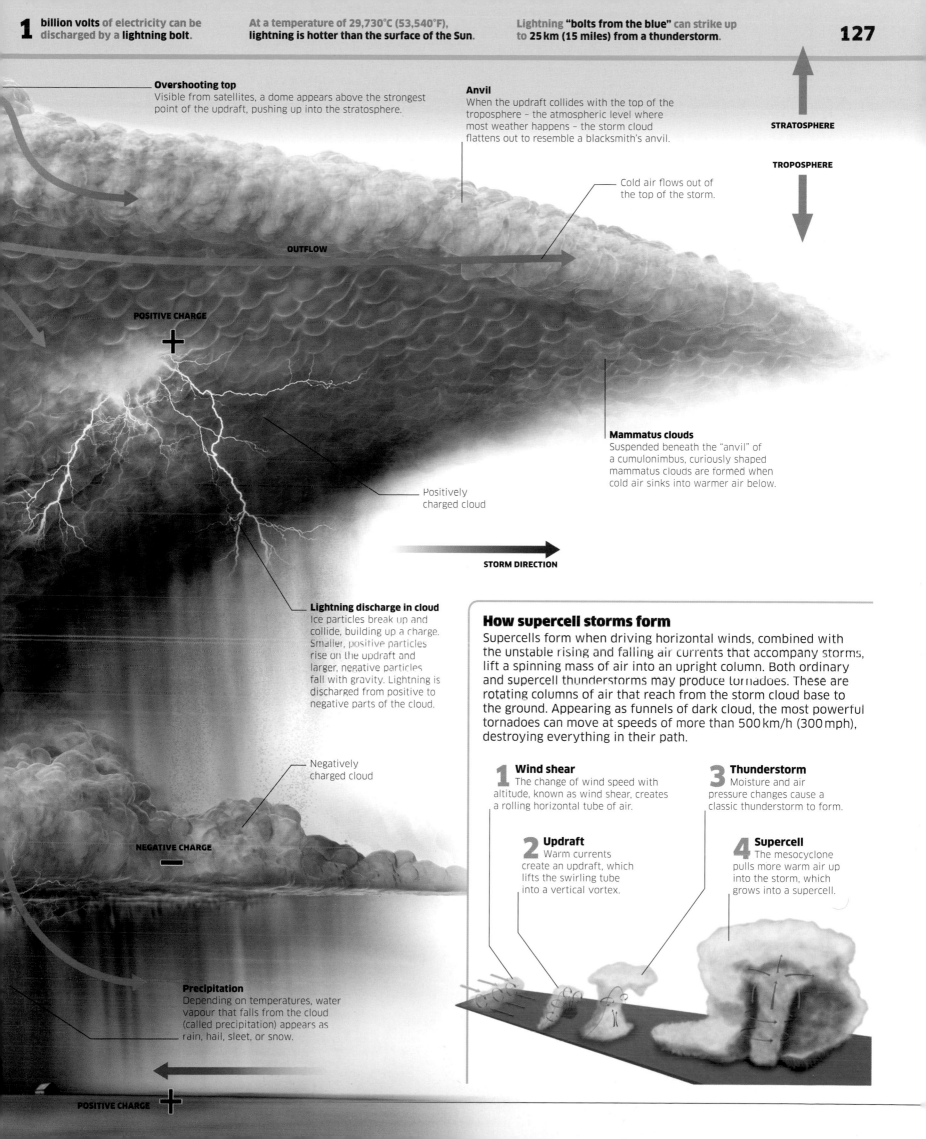

Overshooting top
Visible from satellites, a dome appears above the strongest point of the updraft, pushing up into the stratosphere.

Anvil
When the updraft collides with the top of the troposphere – the atmospheric level where most weather happens – the storm cloud flattens out to resemble a blacksmith's anvil.

STRATOSPHERE

TROPOSPHERE

Cold air flows out of the top of the storm.

OUTFLOW

POSITIVE CHARGE
+

Mammatus clouds
Suspended beneath the "anvil" of a cumulonimbus, curiously shaped mammatus clouds are formed when cold air sinks into warmer air below.

Positively charged cloud

STORM DIRECTION

Lightning discharge in cloud
Ice particles break up and collide, building up a charge. Smaller, positive particles rise on the updraft and larger, negative particles fall with gravity. Lightning is discharged from positive to negative parts of the cloud.

Negatively charged cloud

NEGATIVE CHARGE
—

Precipitation
Depending on temperatures, water vapour that falls from the cloud (called precipitation) appears as rain, hail, sleet, or snow.

POSITIVE CHARGE
+

How supercell storms form

Supercells form when driving horizontal winds, combined with the unstable rising and falling air currents that accompany storms, lift a spinning mass of air into an upright column. Both ordinary and supercell thunderstorms may produce tornadoes. These are rotating columns of air that reach from the storm cloud base to the ground. Appearing as funnels of dark cloud, the most powerful tornadoes can move at speeds of more than 500 km/h (300 mph), destroying everything in their path.

1 Wind shear
The change of wind speed with altitude, known as wind shear, creates a rolling horizontal tube of air.

2 Updraft
Warm currents create an updraft, which lifts the swirling tube into a vertical vortex.

3 Thunderstorm
Moisture and air pressure changes cause a classic thunderstorm to form.

4 Supercell
The mesocyclone pulls more warm air up into the storm, which grows into a supercell.

Climate change

For the last half century, Earth's climate has been getting steadily warmer. The world's climate has always varied naturally, but the evidence suggests that this warming is caused by human activity – and it could have a huge impact on our lives.

Humans make the world warmer mainly by burning fossil fuels such as coal and oil, which fill the air with carbon dioxide that traps the Sun's heat. This is often referred to as global warming, but scientists prefer to talk about climate change because the unpredictable effects include fuelling extreme weather. In future, we can expect more powerful storms and flooding as well as hotter summers and droughts.

Greenhouse effect

The cause of global warning is the greenhouse effect. In the atmosphere, certain gases – known as greenhouse gases – absorb heat radiation that would otherwise escape into space. This causes our planet to be warmer than it would be if it had no atmosphere. The main greenhouses gases are carbon dioxide, methane, nitrous oxide, and water vapour.

2 Reflection
Almost a third of the energy in sunlight is reflected back into space as UV and visible light.

Transport
Petrol- and diesel-guzzling trucks and cars, as well as fuel-burning aeroplanes, produce around 15 per cent of greenhouse gases.

Farming and deforestation
Intensively farmed cows, sheep, and goats release huge amounts of methane, a greenhouse gas. Forests absorb carbon dioxide, so deforestation leaves more carbon dioxide in the atmosphere.

1 Light from the sun
The sunlight that passes through the atmosphere is a mixture of types of radiation: ultraviolet (UV – short wave), visible light (medium wave), and infrared (long wave).

Industry
Heavy industry burning fossil fuels for energy adds about 13 per cent of global greenhouse gas emissions.

Power stations
Burning coal, natural gas, and oil to generate electricity accounts for more than 30 per cent of all polluting carbon dioxide.

3 Absorption
The remaining energy in sunlight is absorbed by the Earth's surface, converted into heat, and emitted into the atmosphere as long-wave, infrared radiation.

8 m (26 ft) – the amount **sea levels** would rise if the **polar ice sheets melted**.

50 per cent – the **increase** in the amount of **carbon dioxide** in the air **since 1980**.

129

9 out of 10
scientists believe that carbon dioxide emissions are the main cause of global warming.

Melting ice caps
Arctic sea ice is melting and the Antarctic ice sheet and mountain glaciers are shrinking fast as the world warms. Melting land ice combined with the expansion of seawater as it warms are raising sea levels. Sea warmth is also adding extra energy into the air, driving storms.

Disappearing ice
The extent of Arctic and Antarctic sea ice shrank to record lows in 2017.

CLIMATE-RELATED DISASTERS
SUCH AS FLOODS, STORMS, AND OTHER
EXTREME WEATHER EVENTS
HAVE INCREASED
THREE TIMES SINCE 1980.

4 Greenhouse trap
Some infrared radiation escapes into space, but some is blocked by greenhouse gases, trapping its warmth in Earth's atmosphere.

Homes
Burning natural gas, oil, coal, and even wood for cooking and to keep homes warm adds almost a tenth of greenhouse gases.

Business
Most of the greenhouse gases generated by business come from electricity use.

Ocean acidification
Carbon dioxide emissions do not only contribute to the greenhouse effect. The gas dissolves in the oceans, making them more acidic. Increasing the acidity of seawater can have a devastating effect on fragile creatures that live in it. It has already caused widespread coral "bleaching", and reefs are dwindling.

LIFE

There is nothing more complex in the entire Universe than living things. Life comes in an extraordinarily diverse range of forms – from microscopic bacteria to giant plants and animals. Each organism has specialized ways of keeping its body working, and interacting with its environment.

1977

Modern times
In the most recent biological developments, things that were thought to be impossible just a hundred years ago became routine. Faulty body parts could be replaced with artificial replicas and even genes could be changed to switch characteristics.

New worlds
American scientists discover deep-sea animals supported by the chemical energy of volcanic vents – the only life not dependent on the Sun and photosynthesis.

1978-1996

New life
The first human "test tube" baby – made with cells fertilized outside the human body – is born in 1978. Then, in 1996, Dolly the sheep becomes the first mammal to be artificially cloned from body cells.

DOLLY THE SHEEP

MODERN TIMES

1960s

Animal behaviour
More biologists begin studying the behaviour of wild animals. In the 1960s, British biologist Jane Goodall discovers that chimpanzees use tools.

1953

The structure of DNA
American and British scientists James Watson and Francis Crick identify that DNA (the genetic code of life packed into cells) has a double helix shape.

1900-1970

Discovering life

Ever since people first began to observe the natural world around them, they have been making discoveries about life and living things.

Biology – the scientific study of life – emerged in the ancient world, when philosophers studied the diversity of life's creatures and medical experts of the day dissected bodies to see how they worked. Hundreds of years later, the invention of the microscope opened up the world of cells and microbes, and allowed scientists to understand the workings of life at the most basic level. At the same time, new insights helped biologists answer some of the biggest questions of all: the causes of disease, and how life reproduces.

1800s

Anaesthetics and antiseptics
The biggest steps in surgery happen in the 1800s: anaesthetics are used to numb pain, while British surgeon Joseph Lister uses antiseptics to reduce infection.

1856-65

Inheritance
An Austrian-born monk called Gregor Mendel, carries out breeding experiments with pea plants that help to explain the inheritance of characteristics.

19th century
The next hundred years saw many of the most important discoveries in biology. Some helped medicine become safer and more effective. Others explained how characteristics are inherited and the evolution of life.

19TH CENTURY

Timeline of discoveries
More than 2,000 years of study and experiment have brought biology into the modern age. While ancient thinkers began by observing the plants and animals around them, scientists today can alter the very structure of life itself.

500 BCE

Antiquity to 16th century
Ancient civilizations in Europe and Asia were the birthplace of science. Here, the biologists of the day described the anatomy (structure) of animals and plants and used their knowledge to invent ways to treat illness.

Describing fossils
Many ancient peoples discover fossils. In the year 500 BCE, Xenophanes, a Greek philosopher, proposes that they are the remains of animals from ancient seas that once covered the land.

Healing theories
Early medical doctors believe that illness is caused by an imbalance of bodily fluids, called humours, that can be treated with bloodsucking leeches.

LEECHES

1315

Early anatomy
Human anatomy is scrutinized in detail by cutting open dead bodies. Dissections are even public spectacles – the first public one is carried out in 1315.

BEFORE 1600

The **Greek philosopher Aristotle** produced the **first classification** of animals and **separated vertebrates from invertebrates**.

In 2001, scientists published the results of the **Human Genome Project**: a catalogue of all human genes.

133

2015

2010s

2017

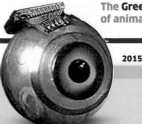

Artificial parts
False limbs had been used since antiquity, but the 20th century brings more sophisticated artificial body parts. The first bionic eye is implanted in 2015.

Fossil evidence
More discoveries of ancient creatures, often preserved in amber, lead scientists to new conclusions, such as the realization that many dinosaurs had feathers.

Changing genes
In the late 20th century scientists become able to edit the genes of living things. In 2017, some mosquitoes are genetically altered to prevent the spread of the disease malaria.

1970 - PRESENT

1930s

1928

1900s

The rise of ecology
The study of ecology (how organisms interact with their surroundings) emerges in the 1930s, as British botanist Arthur Tansley introduces the idea of ecosystems.

Antibiotics discovered
British biologist Alexander Fleming discovers that a substance – penicillin, the first known antibiotic – stops the growth of microbes. Antibiotics are now used to treat many bacterial infections.

Chromosomes and genes
American scientist Thomas Hunt Morgan carries out experiments on fruit flies, which prove that the units of inheritance are carried as genes on chromosomes.

1859

1860s

Early 20th century
Better microscopes and advances in studying the chemical make-up of cells helped to show how all life carries a set of building instructions – in the form of chromosomes and DNA, while ecology and behaviour became new topics of focus.

Evolution
British biologist Charles Darwin publishes a book called *On the Origin of Species*, explaining how life on Earth has evolved by natural selection.

CHARLES DARWIN

Microbes
An experiment by the French biologist Louis Pasteur proves microbes are sources of infection. It also disproves a popular theory which had argued that living organisms could be spontaneously generated from non-living matter.

MICROBE

20TH CENTURY

1800 - 1900

▶

1770s

1735

Vaccines invented
A breakthrough in medicine, the first vaccine is used by British doctor Edward Jenner to protect against a deadly disease – smallpox.

Photosynthesis discovered
In the 1770s, the experiments of a Dutch biologist, Jan Ingenhousz, show that plants need light, water, and carbon dioxide to make sugar.

Classifying life
Swedish botanist Carl Linnaeus devises a way of classifying and naming plants and animals that is still used today.

VAN LEEUWENHOEK'S MICROSCOPE

Microscopic life
British scientist Robert Hooke views cells down a microscope and inspires a Dutchman, Antony van Leeuwenhoek, to invent his own unique version of a microscope.

1600 - 1800

1796

1665

16TH CENTURY

1628

Cataloguing life
The ancient Greeks are the first to try classifying life, but it is not until the 16th century that species are first catalogued in large volumes.

17TH CENTURY

17th-18th centuries
New scientific experiments added to the wealth of knowledge laid down by the first philosophers. This research helped to answer important questions about life's vital processes, such as blood circulation in animals and photosynthesis in plants.

Blood circulation
A British doctor called William Harvey combines observation with experiment to show how the heart pumps blood around the body.

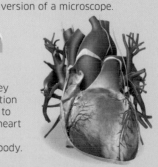

▶

WHAT IS LIFE?

Life can be defined as a combination of seven main actions – known as the characteristics of life – that set living things apart from non-living things. However big or small, every organism must process food, release energy, and excrete its waste. All will also, to some degree, gather information from their surroundings, move, grow, and reproduce.

Life on a leaf

The characteristics of life can all be seen in action on a thumbnail-sized patch of leaf. Tiny insects, called aphids, suck on the leaf's sap and give birth to the next generation, while leaf cells beneath the aphids' feet generate the sap's sugar.

Sensitivity

Sense organs detect changes in an organism's surroundings, such as differences in light or temperature. Each kind of change, or stimulus, is picked up by receptors. With this information, the body can coordinate a suitable response.

Segments near the end of the aphid's antenna contain sense organs.

Antenna
Aphid antennae carry different kinds of sensors, including some that detect odours indicating a leaf is edible.

Nutrition

Food is either consumed or made. Animals, fungi, and many single-celled organisms take food into their body from their surroundings. Plants and algae make food inside their cells, by using light energy from the Sun to convert carbon dioxide and water into sugars and other nutrients.

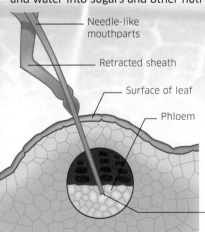

Needle-like mouthparts

Retracted sheath

Surface of leaf

Phloem

Proboscis
Like most other animals, aphids pass food through a digestive system, from which nutrients move into the body's cells. Aphids can only drink liquid sap. They use a sharp proboscis that works like a needle to puncture a leaf vein to get sap.

Pressure in the leaf's vein forces sap up through the proboscis of the aphid.

Movement

Plants are rooted in the ground, but can still move their parts in response to their surroundings, for instance, to move towards a light source. Animals can move their body parts much faster by using muscles, which can even carry their entire body from place to place.

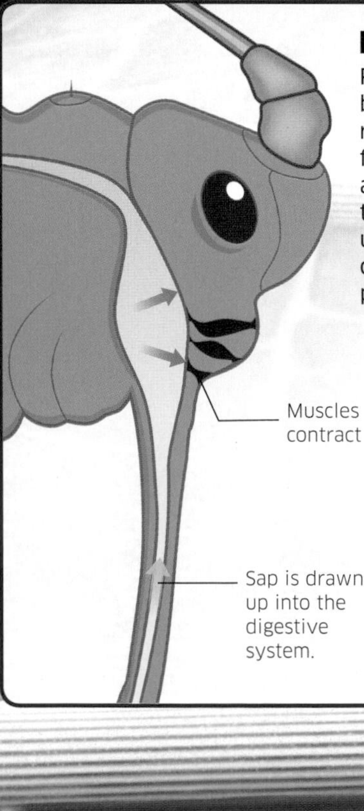

Muscles contract

Head muscles
Muscles are found all over an aphid's body. As the aphid eats, muscles in its head contract (shorten) to pull and widen its feeding tube. This allows it to consume the sap more effectively.

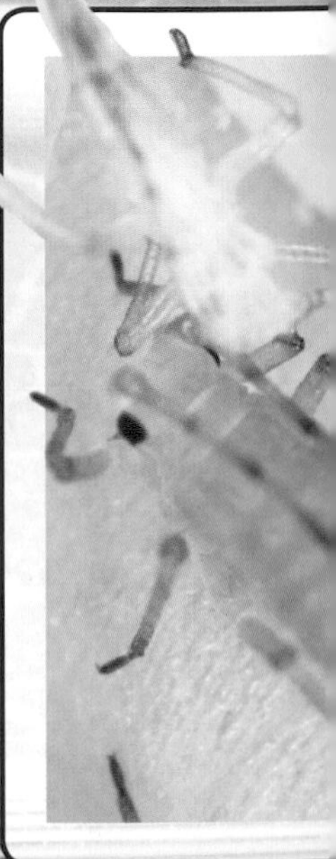

Sap is drawn up into the digestive system.

Birth
A female aphid gives birth to live young.

Reproduction

By producing offspring, organisms ensure that their populations survive, as new babies replace the individuals that die. Breeding for most kinds of organisms involves two parents reproducing sexually by producing sex cells. But some organisms can breed asexually from just one parent.

Babies in babies
Some female aphids carry out a form of asexual reproduction where babies develop from unfertilized eggs inside the mother's body. A further generation of babies can develop inside the unborn aphids.

The daughters that are old enough to be born already contain the aphid's granddaughters.

Respiration

Organisms need energy to power their vital functions, such as growth and movement. A chemical process happens inside the cells to release energy, called respiration. It breaks down certain kinds of foods, such as sugar. Most organisms take in oxygen from the environment to use in their respiration.

Some energy is used to move materials around, including into and out of cells.

Energy is released from food.

FOOD

ENERGY

Some energy is used to build materials inside the cell to help the body grow.

PLANT CELL

Excretion

Hundreds of chemical reactions happen inside living cells, and many of these reactions produce waste substances that would cause harm if they built up. Excretion is the way an organism gets rid of this waste. Animals have excretory organs, such as kidneys, to remove waste, but plants use their leaves for excretion.

Excretion by leaf
Plant leaves have pores, called stomata, for releasing waste gases, such as oxygen and carbon dioxide.

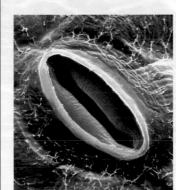

Growth

All organisms get bigger as they get older and grow. Single cells grow very slightly and stay microscopic, but many organisms, such as animals and plants, have bodies made up of many interacting cells. As they grow, these cells divide to produce more cells, making the body bigger.

Moulting
The body of an aphid is covered in a tough outer skin called an exoskeleton. In order to grow, an aphid must periodically shed this skin so its body can get bigger. Its new skin is initially soft and flexible, but soon toughens.

⊙ SEVEN KINGDOMS OF LIFE

Living things are classified into seven main groups called kingdoms. Each kingdom contains a set of organisms that have evolved to perform the characteristics of life in their own way.

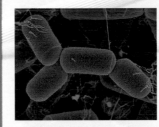

Archaea
Looking similar to bacteria, many of these single-celled organisms survive in very extreme environments, such as hot, acidic pools.

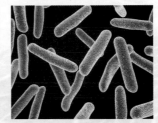

Bacteria
The most abundant organisms on Earth, bacteria are usually single-celled. They either consume food, like animals do, or make it, like plants do.

Algae
Simple relatives of plants, algae make food by photosynthesis. Some are single-celled, but others, such as seaweeds and this *Pandorina*, are multicelled.

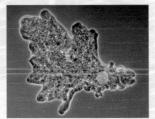

Protozoa
These single-celled organisms are bigger than bacteria. Many of them behave like miniature animals, by eating other microscopic organisms.

Plants
Most plants are anchored to the ground by roots and have leafy shoots to make food by photosynthesis.

Fungi
This kingdom includes toadstools, mushrooms, and yeasts. They absorb food from their surroundings, often by breaking down dead matter.

Animals
From microscopic worms to giant whales, all animals have bodies made up of large numbers of cells, and feed by eating or absorbing food.

The fossil record

Fossils from prehistoric times show just how much life has changed across the ages, and how ancient creatures are related to the organisms on Earth today.

Life has been evolving on our planet for more than four billion years – ever since it was just a world of simple microbes. Across this vast expanse of time, more complex animals and plants developed. Traces of their remains – found as fossils in prehistoric rocks – have helped us to work out their ancestry.

1 Megalosaurus
Theropods, such as *Megalosaurus*, were meat-eating dinosaurs that walked on two legs. Some smaller, feathered theropods were the ancestors of birds.

EARLIER DINOSAUR ANCESTORS

Like most birds, theropods had feet with three forward-pointing toes, and hollow bones.

170 MILLION YEARS AGO

150 MILLION YEARS AGO

The origins of birds

Fossilized skeletons show us that the first prehistoric birds were remarkably similar to a group of upright-walking dinosaurs. From these fossils, it is possible to see how their forelimbs evolved into wings for flight, and how they developed the other characteristics of modern birds.

Archaeopteryx fossil
This near-perfect fossil of an *Archaeopteryx* has been preserved in soft limestone. Around the animal's wing bones, the imprints left by the feathers are clearly visible.

How fossils form

Fossils are the remains or impressions of organisms that died more than 10,000 years ago. Some fossils have recorded what is left of entire bodies, but usually only fragments, such as parts of a bony skeleton, have survived.

Skeletons and other hard parts are more likely to leave an impression than soft tissues.

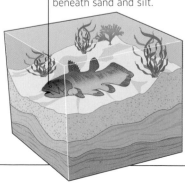

1 Death
Bodies that settled under water or in floodplains could be quickly buried beneath sand and silt.

2 Burial
Layers of sediment cover the body and build up into rock on top of it.

3 Reveal
Millions of years later, movements of Earth's crust cause rocks to move upwards, exposing the fossil on dry land.

Preserved in time
When animals in the prehistoric world died, their bodies were more likely to be preserved if they were quickly buried. Rotting under layers of sediment, the body slowly turned into mineral, until the resulting fossil was exposed by erosion.

Over millions of years, groups of organisms split up as they evolve and become adapted to new environments or situations.

4.2 billion years old – the age of the **oldest fossils** discovered. These were tiny microbes in rocks.

On average, **each species survives for about a million years**, before it **becomes extinct or evolves** into something else.

137

Confuciusornis fossil
Lots of preserved *Confuciusornis* specimens have long tail streamers. These are now thought to be exclusive to males and to have been used in displays to attract a mate during breeding season.

Mass extinction

Life in the prehistoric past was occasionally rocked by catastrophic events that wiped out entire groups of organisms. Five mass extinction events have occurred in the last 500 million years. Many may have been caused by climate change and volcanic eruptions, but there is also strong evidence that the event that eliminated the dinosaurs was caused by an asteroid striking the Earth.

Claws on the bird's thumb and third finger may have helped it climb through trees.

2 Archaeopteryx
Thought to be the first true bird, *Archaeopteryx* had feathered wings, but retained dinosaur characteristics, such as clawed forelimbs, a toothed beak and a bony tail. Its small wing muscles suggest it may not have been able to fly well.

3 Confuciusornis
Thirty million years after *Archaeopteryx*, *Confuciusornis* emerged. It had a tail of feathers that lacked a bony support, and a toothless beak. Its flight feathers were longer than *Archaeopteryx*, but it still could not flap as well as modern birds.

4 Ichthyornis
Living just before the extinction of the dinosaurs, *Ichthyornis* resembled a modern seabird, and was about the size of a gull. Although it had strong flight muscles, its bill still contained sharp teeth that helped it catch fish.

120 MILLION YEARS AGO

90 MILLION YEARS AGO

Ichthyornis had a well-developed breastbone for supporting strong flight muscles.

A mass extinction 66 million years ago drove dinosaurs to extinction, but their bird descendants survived.

PRESENT DAY

5 Ruby-throated hummingbird
A lightweight skeleton and strong muscles help most modern, toothless birds fly with far greater skill than any of their ancestors.

Evolution

All living things are related and united by a process called evolution. Over millions of years, evolution has produced all the species that have ever lived.

Change is a fact of life. Every organism goes through a transformation as it develops and gets older. But over much longer periods of time – millions or billions of years – entire populations of plants, animals, and microbes also change by evolving. All the kinds of organisms alive today have descended from different ones that lived in the past, as tiny variations throughout history have combined to produce entirely new species.

Song thrushes are an important predator of snails, often foraging in bushes and trees to find prey.

The bird smashes a snail on a hard stone to get to the soft body inside.

Natural selection

The characteristics of living things are determined by genes (see pp.180–181), which sometimes change as they are passed down through generations – producing mutations. All the variety in the natural world – such as the colours of snail shells – comes from chance mutations, but not all of the resulting organisms do well in their environments. Only some survive to pass their attributes on to future generations – winning the struggle of natural selection.

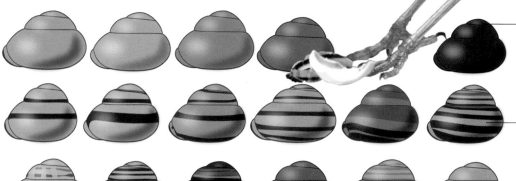

Shells of grove snails vary in colour from yellow to dark brown, depending upon the genes they carry.

Some snail genes cause their shells to develop banding patterns.

Dry grassy habitat
Against a background of dry grass, snails with darker shells are most easily spotted, causing the paler ones to survive in greater numbers.

The song thrush hunts by sight, so picks out the most visible snails.

Dark woodland habitat
In woodland, grove snails with shells that match the dark brown leaf litter of the woodland floor are camouflaged and survive, but yellow-shelled snails are spotted by birds.

Brown snails are more likely to survive in woodland, so more will build up in this area over time.

Hedgerow habitat
In some sun-dappled habitats with a mixture of grass, twigs, and leaves, stripy-shelled grove snails are better disguised, and plain brown or yellow ones become prey.

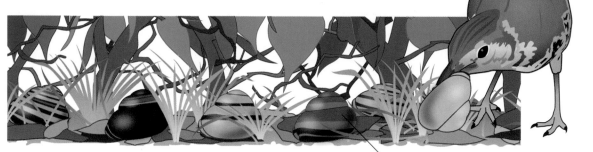

Stripy shells break up the outline of the snails, so they are not easily seen.

Africa's **Lake Malawi** contains more than **500 species of cichlid fishes** that have **evolved from a single ancestral fish** within the last million years.

50 years is all the time it takes for **some infectious bacteria** to **evolve resistance** to antibiotic drugs.

139

How new species emerge

Over a long period of evolution, varieties of animals can end up becoming so different that they turn into entirely new species – a process called speciation. This usually happens when groups evolve differences that stop them from breeding outside their group, especially when their surroundings change so dramatically that they become physically separated from others.

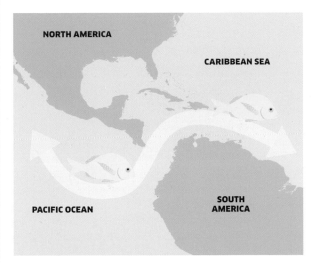

1 Ancestral species
Five million years ago, before North and South America were joined, a broad sea channel swept between the Pacific Ocean in the west and the Caribbean in the east. Marine animals, such as the reef-dwelling porkfish, could easily mix with one another in the open waters. Porkfish from western and eastern populations had similar characteristics and all of them could breed together, so they all belonged to the same species.

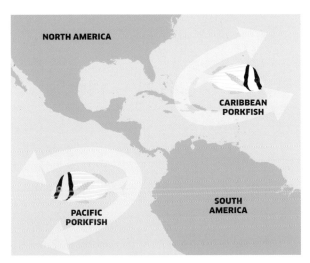

2 Modern species
The shifting of Earth's crust caused North and South America to collide nearly 3 million years ago. This cut off the sea channel, isolating populations of porkfish on either side of Central America. Since then, the two populations have evolved so differently that they can no longer breed with each other. Although they still share a common ancestor, today the whiter Caribbean porkfish and the yellower Pacific porkfish are different species.

Evolution on islands

Isolated islands often play host to the most dramatic evolution of all. Animals and plants can only reach them by crossing vast expanses of water, and – once there – evolve quickly in the new and separate environment. This can lead to some unusual creatures developing – such as flightless birds and giant tortoises.

Out of all the reptiles and land mammals of the Galápagos Islands,

97 per cent are found nowhere else in the world.

Tortoise travels
The famous giant tortoises unique to the Galápagos Islands are descended from smaller tortoises that floated there from nearby South America.

Adaptation

Living things that survive the gruelling process of natural selection are left with characteristics that make them best suited to their surroundings. This can be seen in groups of closely related species that live in very different habitats – such as these seven species of bears.

Polar bear
The biggest, most carnivorous species of bear is adapted to the icy Arctic habitat. It lives on fat-rich seal meat and is protected from the bitter cold by a thick fur coat.

Brown bear
The closest relative of the polar bear lives further south in cool forests and grassland. As well as preying on animals, it supplements its diet with berries and shoots.

Black bear
The North American black bear is the most omnivorous species of bear, eating equal amounts of animal and plant matter. This smaller, nimbler bear can climb trees to get food.

Sun bear
The smallest bear lives in tropical Asia and has a thin coat of fur to prevent it from overheating. It has a very sweet tooth and extracts honey from bee hives with its long tongue.

Sloth bear
This shaggy-coated bear from India is adapted to eat insects. It has poorly developed teeth and, instead, relies on long claws and a long lower lip to obtain and eat its prey.

Spectacled bear
The only bear in South America has a short muzzle and teeth adapted for grinding tough plants. It feeds mainly on leaves, tree bark, and fruit, only occasionally eating meat.

Giant panda
The strangest bear of all comes from the cool mountain forests of China. It is almost entirely vegetarian, with paws designed for grasping tough bamboo shoots.

Miniature life

Some organisms are so tiny that thousands of them can live out their lives in a single drop of water.

The minuscule home of the microbe, or microorganism, is a place where sand grains are like giant boulders and the slightest breeze feels like a hurricane. These living things can only be seen through a microscope, but manage to find everything they need to thrive in soil, oceans, or even deep inside the bodies of bigger animals.

GIARDIA
Kingdom: Protozoa

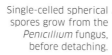

Animal-like microbes that are single-celled are called protozoans. Some, such as amoebas, use extensions of cytoplasm (cell material) to creep along. Others, such as giardia, swim, and absorb their food by living in the intestines of animals.

1/100 mm

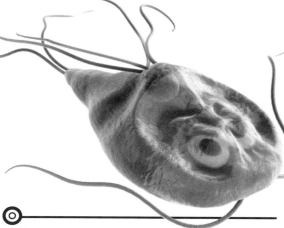

DIATOM
Kingdom: Algae

The biggest algae grow as giant seaweeds, but many, such as diatoms, are microscopic single cells. All make food by photosynthesis, forming the bottom of many underwater food chains that support countless lives.

1/100 mm

Diatoms are surrounded by a large cell wall.

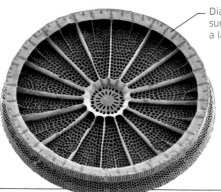

SPIROCHAETE
Kingdom: Bacteria

Any place good for life can be home to bacteria – the most abundant kinds of microorganisms on the planet. They are vital for recycling nutrients, although some – such as the corkscrew-shaped spirochaetes – are parasites that cause disease in humans and other animals.

1/100 mm

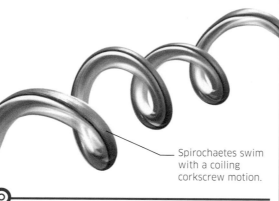

Spirochaetes swim with a coiling corkscrew motion.

PENICILLIUM
Kingdom: Fungi

The microscopic filaments of fungi smother dead material, such as leaf litter, so their digestive juices can break it down. When their food runs out, they scatter dust-like spores, which grow into new fungi.

1/10 mm

Single-celled spherical spores grow from the *Penicillium* fungus, before detaching.

WATERMEAL
Kingdom: Plants

The smallest plant, called watermeal, floats on ponds, blanketing the surface in its millions. A hundred could sit comfortably on a fingertip, each one carrying a tiny flower that allows it to reproduce.

1 mm

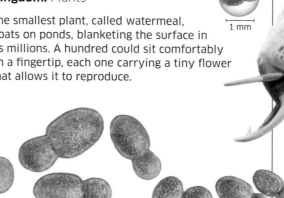

THERMOPLASMA
Kingdom: Archaea

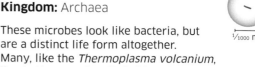

These microbes look like bacteria, but are a distinct life form altogether. Many, like the *Thermoplasma volcanium*, live in the most hostile habitats imaginable, such as hot pools of concentrated acid.

1/1000 mm

Like bacteria, archaea have no cell nucleus and are protected by a tough cell wall.

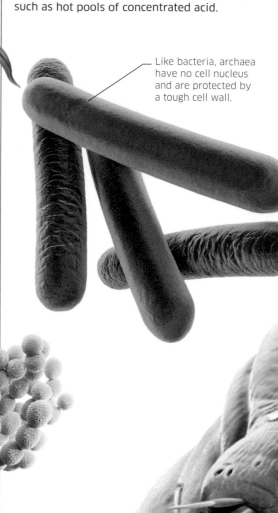

The number of **bacteria in your mouth** is greater than **the number of people on Earth.**

Single-celled **microbes were the first life on Earth** – 4 billion years ago.

141

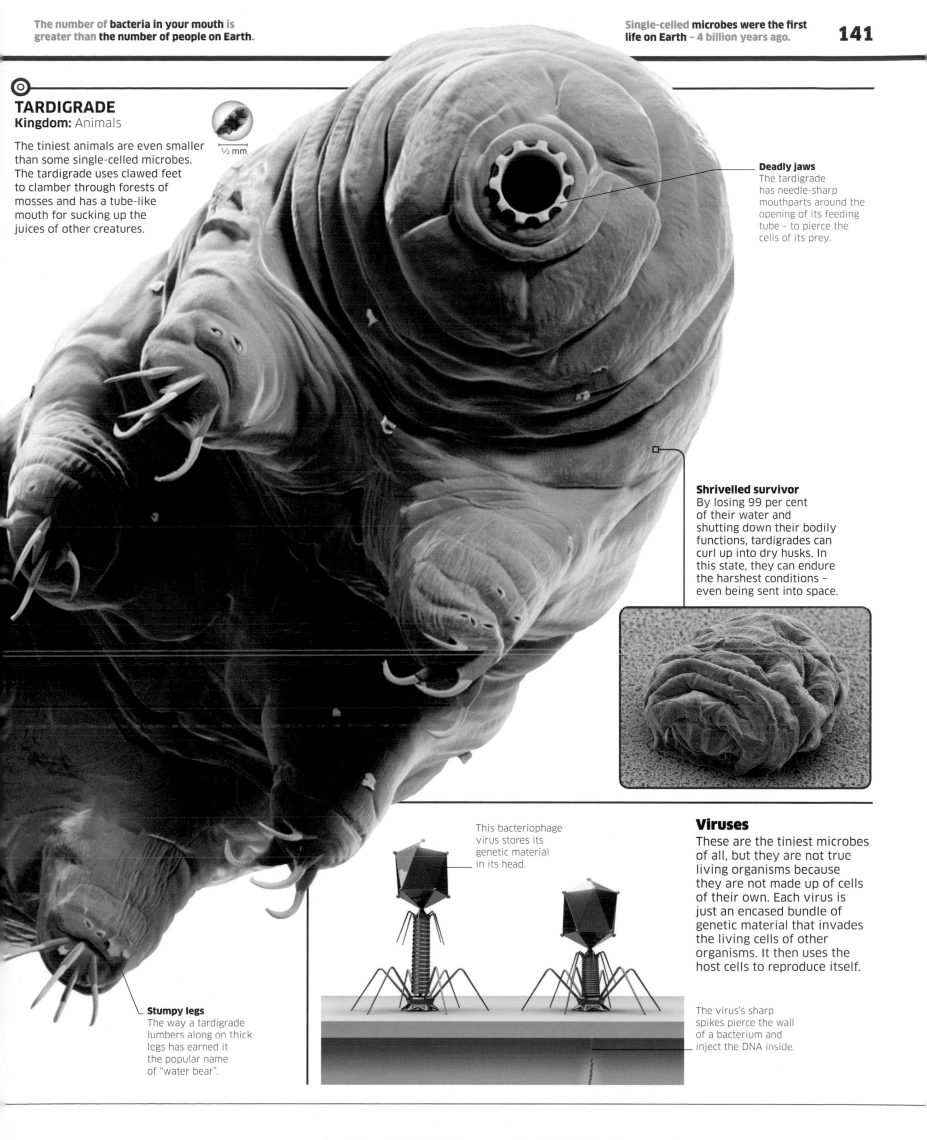

TARDIGRADE
Kingdom: Animals

The tiniest animals are even smaller than some single-celled microbes. The tardigrade uses clawed feet to clamber through forests of mosses and has a tube-like mouth for sucking up the juices of other creatures.

½ mm

Deadly jaws
The tardigrade has needle-sharp mouthparts around the opening of its feeding tube – to pierce the cells of its prey.

Shrivelled survivor
By losing 99 per cent of their water and shutting down their bodily functions, tardigrades can curl up into dry husks. In this state, they can endure the harshest conditions – even being sent into space.

Stumpy legs
The way a tardigrade lumbers along on thick legs has earned it the popular name of "water bear".

This bacteriophage virus stores its genetic material in its head.

Viruses
These are the tiniest microbes of all, but they are not true living organisms because they are not made up of cells of their own. Each virus is just an encased bundle of genetic material that invades the living cells of other organisms. It then uses the host cells to reproduce itself.

The virus's sharp spikes pierce the wall of a bacterium and inject the DNA inside.

Cells

The living building blocks of animals and plants, cells are the smallest units of life. Even at this microscopic level, each one contains many complex and specialized parts.

Cells need to be complex to perform all the jobs needed for life. They process food, release energy, respond to their surroundings, and – within their minuscule limits – build materials to grow. In different parts of the body, many cells are highly specialized. Cells in the muscles of animals can twitch to move limbs and those in blood are ready to fight infection.

Centriole
Structural proteins called microtubules are assembled around a cylindrical arrangement known as a centriole.

Golgi apparatus
The Golgi apparatus packages proteins and sends them to where they are needed.

Cytoplasm
The jelly-like cytoplasm holds all the cell's parts – known as organelles.

Cell membrane
A thin, oily layer controls the movement of substances into and out of the cell.

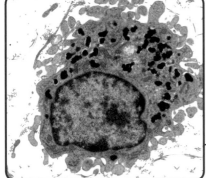

Nucleus
The nucleus (dark purple) controls the activity of the cell. It is packed with DNA (deoxyribonucleic acid) – the cell's genetic material.

Pseudopodium
One of many finger-like extensions of cytoplasm helps this kind of cell to engulf bacteria.

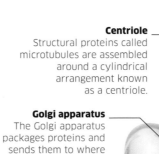

Cells eating cells
A white blood cell is one of the busiest cells in a human body, part of a miniature army that destroys potentially harmful bacteria. Many white blood cells do this by changing shape to swallow invading cells: they extend fingers of cytoplasm that sweep bacteria into sacs for digestion.

1 Bacterium approaches
Bacterial cells are 100 times smaller than blood cells, but potentially cause disease. It is a white blood cell's job to prevent them from invading the body.

2 Food vacuole forms
The blood cell envelops bacteria within its cytoplasm, trapping them in fluid-filled sacs called food vacuoles.

3 Digestion begins
Tiny bags of digestive fluid – called lysosomes – fuse with the food vacuole and empty their contents onto the entrapped bacteria.

Bacteria cells look different from those of plants and animals: they **do not contain a nucleus**, mitochondria, or chloroplasts.

143

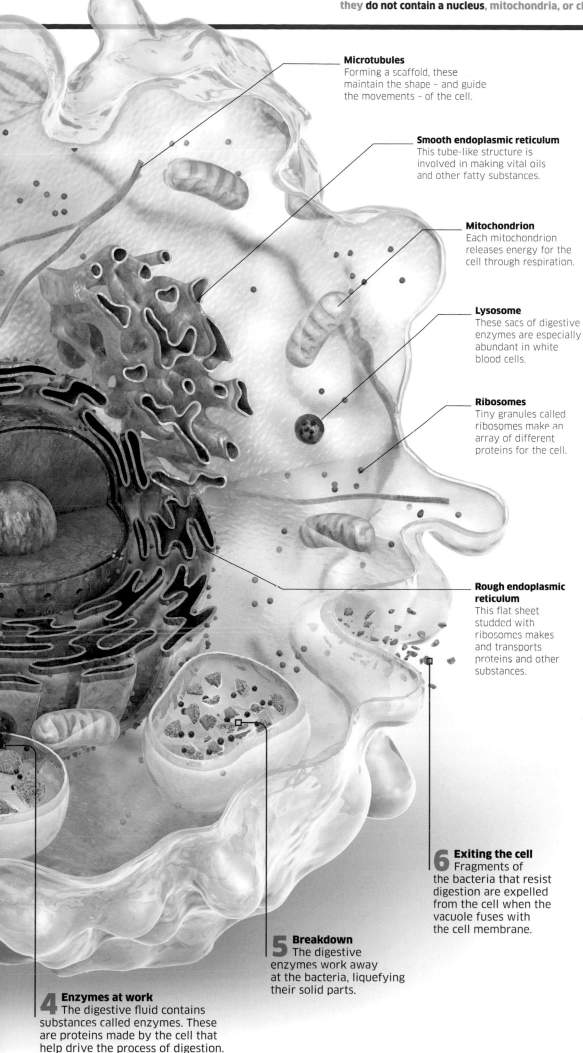

Microtubules
Forming a scaffold, these maintain the shape – and guide the movements – of the cell.

Smooth endoplasmic reticulum
This tube-like structure is involved in making vital oils and other fatty substances.

Mitochondrion
Each mitochondrion releases energy for the cell through respiration.

Lysosome
These sacs of digestive enzymes are especially abundant in white blood cells.

Ribosomes
Tiny granules called ribosomes make an array of different proteins for the cell.

Rough endoplasmic reticulum
This flat sheet studded with ribosomes makes and transports proteins and other substances.

6 Exiting the cell
Fragments of the bacteria that resist digestion are expelled from the cell when the vacuole fuses with the cell membrane.

5 Breakdown
The digestive enzymes work away at the bacteria, liquefying their solid parts.

4 Enzymes at work
The digestive fluid contains substances called enzymes. These are proteins made by the cell that help drive the process of digestion.

Enzymes

Cells make complex molecules called proteins, many of which work as enzymes. Enzymes are catalysts – substances that increase the rate of chemical reactions, and can be used again and again. Each type of reaction needs a specific kind of enzyme.

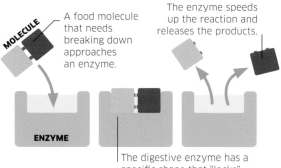

A food molecule that needs breaking down approaches an enzyme.

The enzyme speeds up the reaction and releases the products.

The digestive enzyme has a specific shape that "locks" onto the food molecule.

Cell variety

Unlike animal cells, plant cells are ringed by a tough cell wall and many have food-making chloroplasts. Both animals and plants have many specialised cells for different tasks.

ANIMAL CELLS

Fat cell
Its large droplet of stored fat provides energy when needed.

Bone-making cell
Long strands of cytoplasm help this cell connect to others.

Ciliated cell
Hair-like cilia waft particles away from airways.

Secretory cell
These cells release useful substances, such as hormones.

PLANT CELLS

Starch-storing cell
Some root cells store many granules of energy-rich starch.

Leaf cell
Inside this cell, green chloroplasts make food for the plant.

Supporting cell
Thick-walled cells in the stem help support plants.

Fruit cell
Its large sap-filled vacuole helps to make a fruit juicy.

1.5 kg (3.3 lb) – the **weight of the liver**, which is the **largest internal organ** in the human body.

Skeletal system
Some of the hardest parts of the body make up the skeleton. Bone contains living cells but is also packed with hard minerals. This helps it support the stresses and strains of the moving body, and to protect soft organs too.

Circulatory system
No living cell is very far from a blood vessel. The circulatory system serves as a lifeline to every cell. It circulates food, oxygen, and chemical triggers – such as hormones – as well as transporting waste to the excretory organs.

Digestive system
By processing incoming food, the digestive system is the source of fuel and nourishment for the entire body. It breaks down food to release nutrients, which then seep into the bloodstream to be circulated to all living cells.

Muscular system
The moving parts of the body rely on muscles that contract when triggered to do so by a nerve impulse or chemical trigger. Contraction shortens the muscle, which pulls on a part of the body to cause motion.

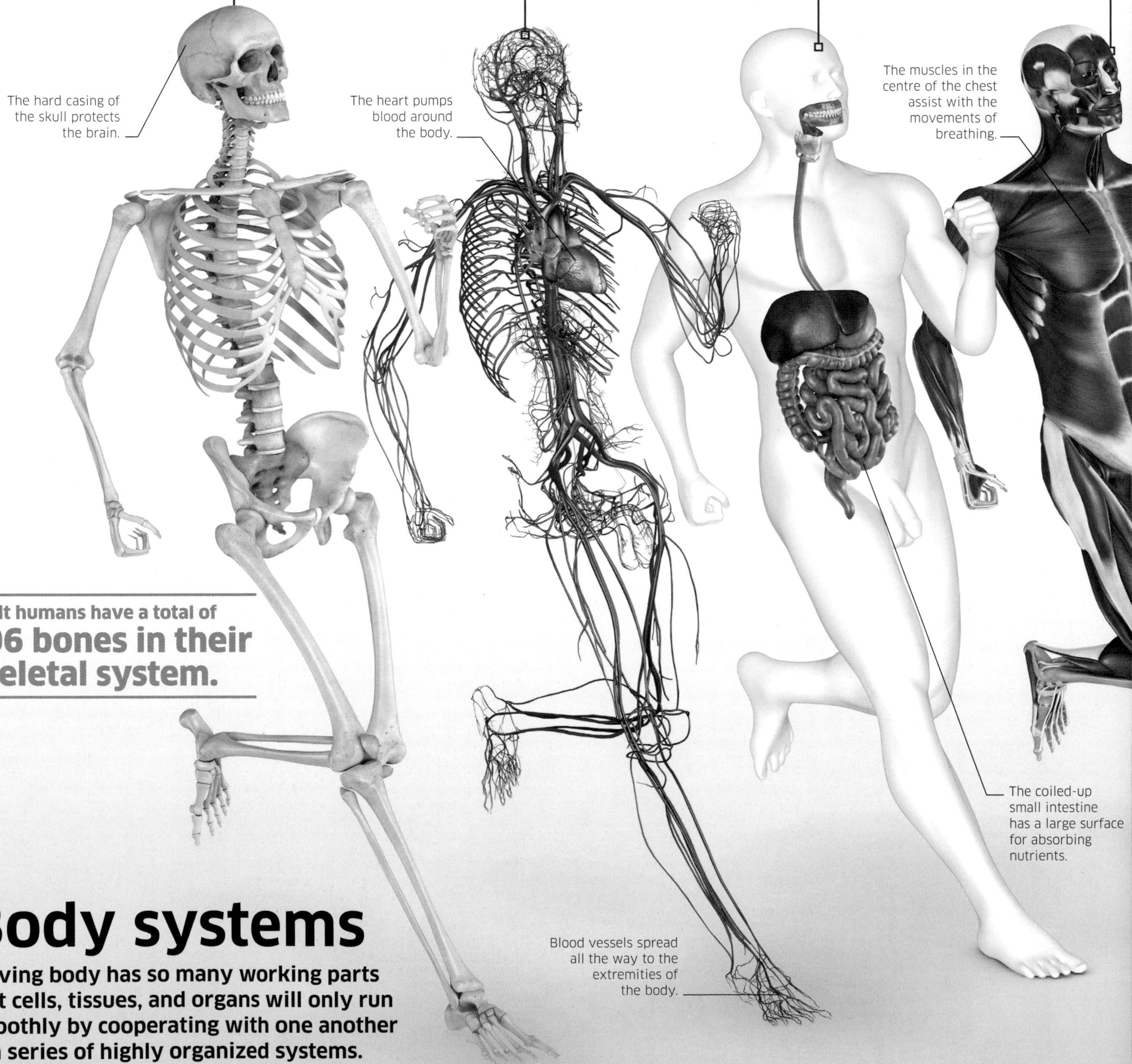
The hard casing of the skull protects the brain.
The heart pumps blood around the body.
The muscles in the centre of the chest assist with the movements of breathing.
The coiled-up small intestine has a large surface for absorbing nutrients.
Blood vessels spread all the way to the extremities of the body.

Adult humans have a total of
206 bones in their skeletal system.

Body systems

A living body has so many working parts that cells, tissues, and organs will only run smoothly by cooperating with one another in a series of highly organized systems.

Each system is designed to carry out a particular function essential to life – whether breathing, eating, or reproducing. Just as organs are interconnected in organ systems, the systems interact, and some organs, such as the pancreas, even belong to more than one system.

Human body systems
There are 12 systems of the human body, of which eight of the most vital are shown here. The others are the urinary system (see pp.162–163), the integumentary system (skin, hair, and nails), the lymphatic system (which drains excess fluid), and the endocrine system (which produces hormones).

Blood is a liquid tissue and contains more than **4 million red blood cells** per cubic millimetre – **the most abundant kind of cell** in the human body.

The skin is the largest organ of all – accounting for more than **10 per cent of total human body weight.**

145

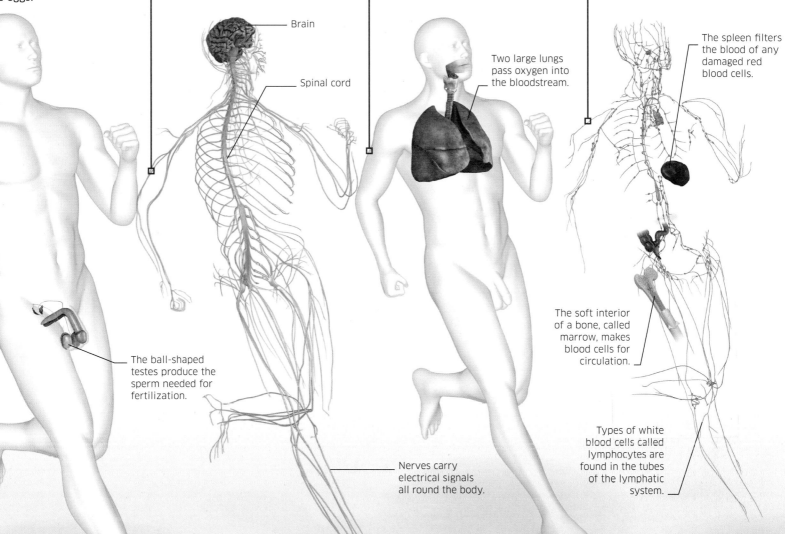

Reproductive system
Differing significantly between the sexes, the reproductive system produces the next generation of life. Female organs produce eggs, and the female body also hosts the developing child before it is born. Male organs produce sperm to fertilize the eggs.

The ball-shaped testes produce the sperm needed for fertilization.

Nervous system
A network of nerves carries high-speed electrical impulses all around the body. These are coordinated by the brain and the spinal cord. When they reach their destination, they trigger responses that control the body's behaviour.

Brain

Spinal cord

Nerves carry electrical signals all round the body.

Respiratory system
The lungs in the respiratory system breathe in air and extract oxygen from it. This oxygen is used in cells to release energy to power the body, while waste carbon dioxide from this reaction is expelled back out through the nose and mouth.

Two large lungs pass oxygen into the bloodstream.

Immune system
The immune system is made up of white blood cells. These travel around the body in the circulatory and lymphatic systems, as well as being found in certain tissues. They help to fight off infectious microbes that have invaded the body.

The spleen filters the blood of any damaged red blood cells.

The soft interior of a bone, called marrow, makes blood cells for circulation.

Types of white blood cells called lymphocytes are found in the tubes of the lymphatic system.

Building a body
Each of the trillions of cells that make up a human body are busy with life's vital processes, such as processing food. But cells are also organized for extra tasks in arrangements called tissues, such as muscles and blood. Multiple tissues, in turn, make up organs, each of which has a specific vital function. A collection of organs working together to carry out one process is called a system.

Cell
The basic building blocks of life, cells can be specialized for a variety of different tasks.

Tissue
Groups of complementary cells work together in tissues that perform particular functions.

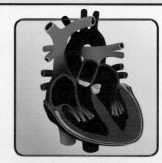

Organ
Combinations of tissues are assembled together to make up organs, such as the human heart.

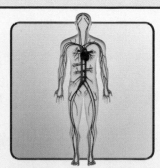

System
Complementary organs are connected into organ systems, which carry out key body processes.

Photosynthesizers
Leaves contain a green pigment called chlorophyll. This traps the energy of sunlight, which is used to build sugars. The process, called photosynthesis, is the origin of virtually all the food chains on Earth.

Flies are drawn to the giant Rafflesia flower because it has the odour of their favourite food: rotting meat.

An indigo flycatcher snatches flies attracted to the foul stench of the Rafflesia flower.

Nutrition

All life needs food – whether it's the sugary sap made in the green leaves of plants, or the solid meals eaten by hungry animals.

Food gives organisms the fuel to power all the living processes that demand energy, such as growth. Animals, fungi, and many microbes consume it from their surroundings – by eating or absorbing the materials of other organisms, living or dead. In contrast, plants and other microbes start with very simple chemical ingredients, such as carbon dioxide and water, and use these to make food inside their cells.

A Kinabalu pit viper hunts small mammals and birds.

What is food?

The nutrients in food come from a complex mixture of molecules – each one containing carbon, hydrogen, and oxygen as its main elements. Three main groups – carbohydrates, fats, and proteins – make up the bulk of food molecules, although all organisms require different amounts of each type.

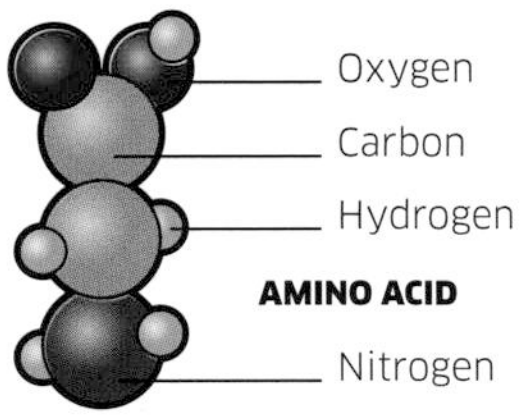

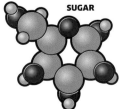

Oxygen
Carbon
Hydrogen
AMINO ACID
Nitrogen

FATTY ACID

Proteins
Groups of atoms called amino acids link into chains of proteins, which help with growth and repair.

SUGAR

Carbohydrates
Rings of atoms called sugars provide energy and link to form chains of starch.

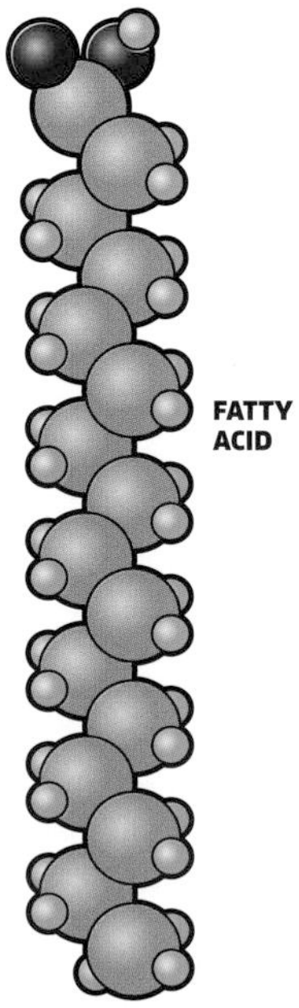

Fats and oils
Used for storing energy or building cells, these are made of long molecules called fatty acids.

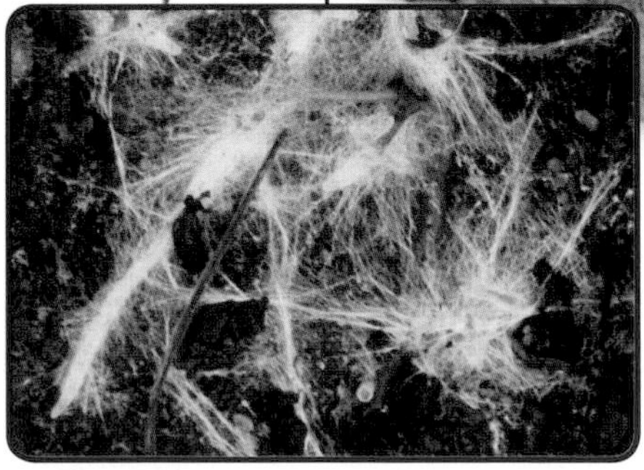

Mycorrhizae
A network of fungus filaments – called mycorrhizae – grows among plant roots. Together, roots and filaments have a feeding partnership: the plants pass sugars to the fungi in exchange for minerals gathered by the fungi.

Predators
Animals that prey on others are called predators. Leeches are famous for sucking blood, but the giant red leech has a taste for meat – grabbing giant earthworms as they emerge from burrows after rainfall.

Tropical rainforests produce nearly
40 billion tonnes of food each year.

147

Parasites

Surprisingly, the world's biggest flower
is produced by a plant with no leaves.
Rafflesia's massive bloom stinks of rotting
meat to attract pollinating blowflies, but
the rest of the plant grows as spreading
tissue inside a tropical vine. A parasite,
it steals food from the vine because it
cannot photosynthesize for itself.

Hotbed of nutrition

A rainforest floor in Borneo is a busy
community of living things, all striving
for nourishment. While green-leaved
plants make the food upon which,
ultimately, everything else depends, a
multitude of predators, parasites, and
decomposers are fed by living prey
and an abundance of dead matter.

Mountain tree
shrews nourish the
pitchers with their
droppings – and are
rewarded with a
lick of sweet nectar.

Insectivorous plants

Where the soil is low in
certain minerals, some plants
seek other sources of food.
The leaves of pitcher plants
develop into vessels that
contain pools of fluid for
digesting drowning insects
and even the droppings
of occasional mammals.

Saprophytes

Toadstools and other fungi
are saprophytes – meaning
that they absorb the liquefied
remains of dead matter. They
are made up of microscopic
filaments, called hyphae,
that penetrate the soil
and cling to dead matter,
simultaneously releasing
digestive juices and soaking
up the digested products.

Soil contains dead matter,
which releases minerals
into the ground as
it decomposes.

Bacteria

Most kinds of bacteria digest
dead matter, driving the process
of decomposition. Others process
the chemical energy in minerals
to make their own food and, in
doing so, release nitrates – an
important source of nitrogen
sucked up by plant roots.

Many detritus-eating animals
burrow in soil, where they
are surrounded by their food.

Detritivores

A forest floor is littered with
organic detritus (waste), such
as dead leaves. This provides
abundant food for detritivores,
such as giant blue earthworms,
that have the digestive systems
to cope with this tough material.

Some kinds of ocean algae use brown or red pigment for their photosynthesis, to **make better use of the light wavelengths** that penetrate the water.

Photosynthesis

Virtually all food chains on Earth begin with photosynthesis – the chemical process in green leaves and algae that is critical for making food.

All around the planet when the Sun shines, trillions of microscopic chemical factories called chloroplasts generate enough food to support all the world's vegetation. These vital granules are packed inside the cells of plant leaves and ocean algae. They contain a pigment, called chlorophyll, that makes our planet green and absorbs the Sun's energy to change carbon dioxide and water into life-giving sugar.

Nucleus
Tightly-packed chloroplasts

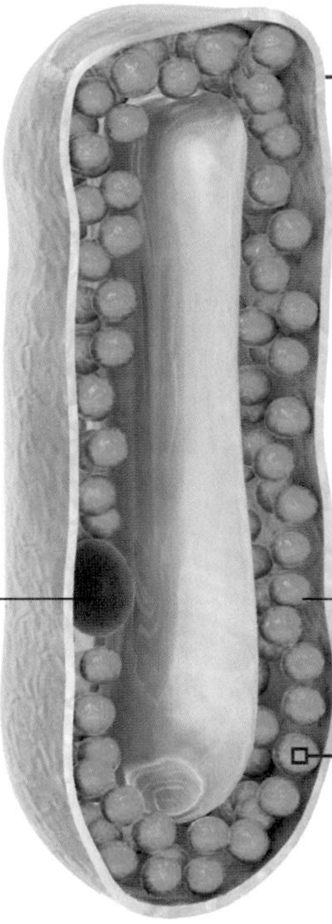

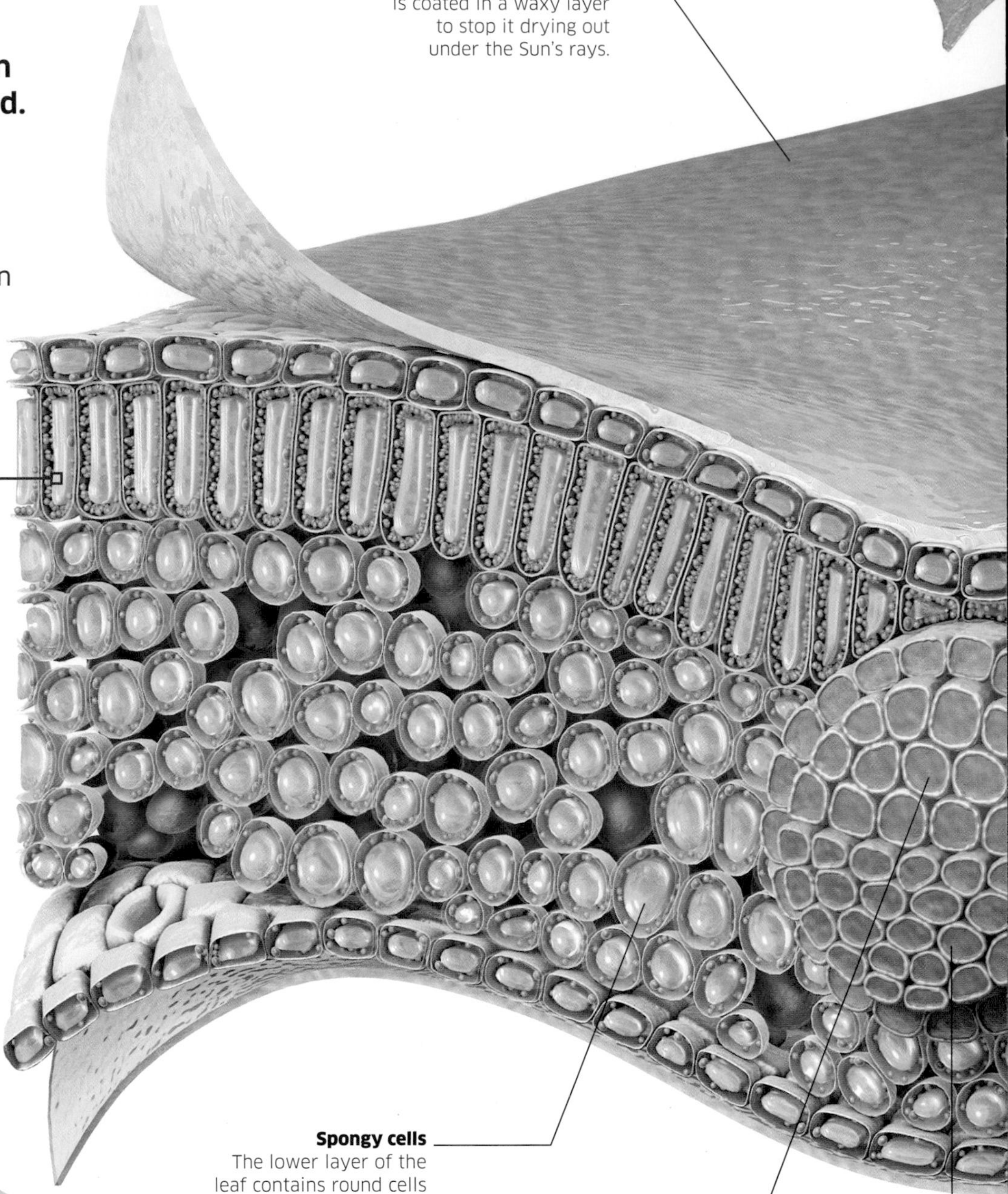

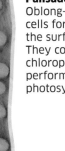

Palisade cell
Oblong-shaped palisade cells form a layer near the surface of the leaf. They contain the most chloroplasts and so perform the most photosynthesis.

Waxy layer
The surface of the leaf is coated in a waxy layer to stop it drying out under the Sun's rays.

Spongy cells
The lower layer of the leaf contains round cells surrounded by air-filled spaces. These spaces help carbon dioxide in the air reach photosynthesizing cells.

Xylem
Tubes called xylem carry water into the leaf.

Phloem
Phloem tubes transport the food made during photosynthesis to other parts of the plant.

Chloroplast
A chloroplast is a bean-shaped granule. Together, all the chloroplasts contain so much of the pigment chlorophyll that the entire leaf appears green.

Fluid around the discs contains chemicals called enzymes that drive the production of sugar.

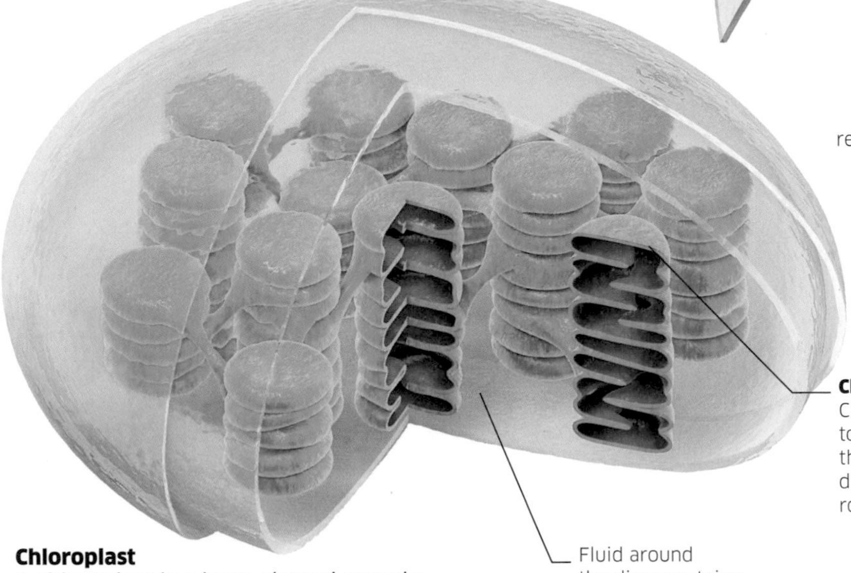

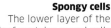

Chlorophyll
Chlorophyll is fixed to membranes around the discs. Having lots of discs means there is more room for chlorophyll.

Inside a leaf

Cells that are near the sun-lit surface of a leaf contain the most chloroplasts. Each chloroplast is sealed by transparent oily membranes and encloses stacks of interconnected discs that are at the heart of the photosynthesis process. The discs are covered in green chlorophyll, which traps light energy from the Sun. This energy then drives chemical reactions that form sugar in the fluid surrounding the discs.

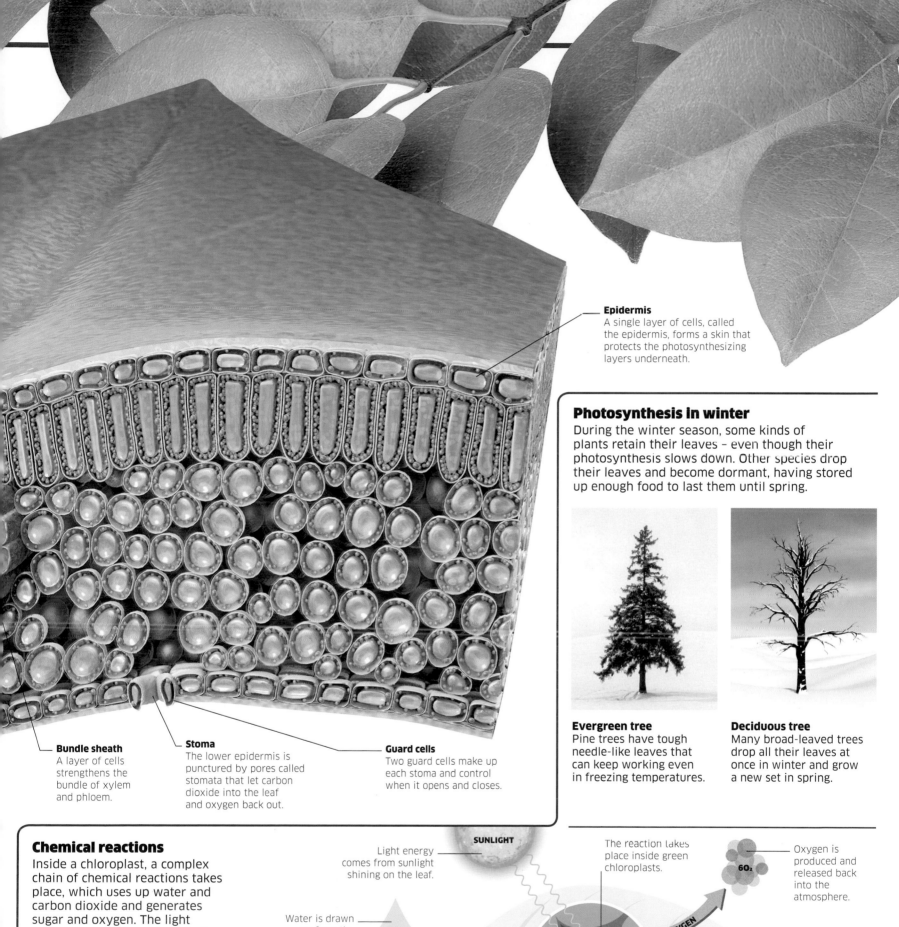

Epidermis
A single layer of cells, called the epidermis, forms a skin that protects the photosynthesizing layers underneath.

Bundle sheath
A layer of cells strengthens the bundle of xylem and phloem.

Stoma
The lower epidermis is punctured by pores called stomata that let carbon dioxide into the leaf and oxygen back out.

Guard cells
Two guard cells make up each stoma and control when it opens and closes.

Photosynthesis in winter
During the winter season, some kinds of plants retain their leaves – even though their photosynthesis slows down. Other species drop their leaves and become dormant, having stored up enough food to last them until spring.

Evergreen tree
Pine trees have tough needle-like leaves that can keep working even in freezing temperatures.

Deciduous tree
Many broad-leaved trees drop all their leaves at once in winter and grow a new set in spring.

Chemical reactions
Inside a chloroplast, a complex chain of chemical reactions takes place, which uses up water and carbon dioxide and generates sugar and oxygen. The light energy trapped by chlorophyll is first used to extract hydrogen from water, and expel the excess oxygen into the atmosphere. The hydrogen is then combined with carbon dioxide to make a kind of sugar called glucose. This provides the energy the plant needs for all the functions of life.

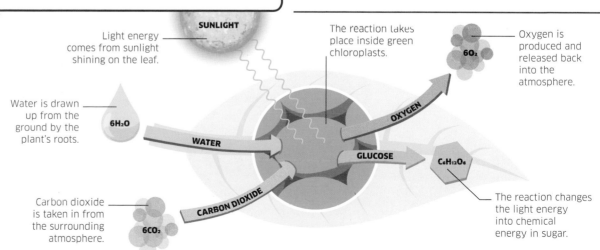

Light energy comes from sunlight shining on the leaf.

SUNLIGHT

The reaction takes place inside green chloroplasts.

Oxygen is produced and released back into the atmosphere.

$6O_2$

OXYGEN

Water is drawn up from the ground by the plant's roots.

$6H_2O$

WATER

GLUCOSE

$C_6H_{12}O_6$

CARBON DIOXIDE

Carbon dioxide is taken in from the surrounding atmosphere.

$6CO_2$

The reaction changes the light energy into chemical energy in sugar.

Tapeworms are **parasites** that live inside the bodies of other animals, and **absorb food without using a digestive system** of their own.

COLORADO BEETLE
Strategy: Leaf eater

Leaves can be a bountiful source of food, but leaf eaters must first get past a plant's defences. Many are specialized to deal with particular plants, such as the Colorado beetle, which eats potato plant leaves that are poisonous to other animals.

VAMPIRE BAT
Strategy: Parasite

Some animals obtain food directly from living hosts – without killing them. Blood suckers, such as the vampire bat, get a meal rich in protein. The bat attacks at night, and is so stealthy that the sleeping victim scarcely feels its bites.

HAGFISH
Strategy: Scavenger

Deep-sea hagfishes are scavengers: they feed on dead matter. By tying themselves into knots, they are able to brace themselves against the carcasses of dead whales so that their spiny jawless mouths can rasp away at the flesh.

COCONUT CRAB
Strategy: Fruit and seed eater

Although many fruits and seeds are packed with nutrients, not all are easily accessible. The world's biggest land crab feasts on coconuts – tough "stone fruits" that its powerful claws must force open to reach the flesh inside.

Feeding strategies

All animals need food to keep them alive – in the form of other organisms, such as plants and animals. Many will go to extreme lengths to obtain their nutrients.

Whether they are plant-eating herbivores, meat-eating carnivores, or omnivores that eat many different foods, all animals are adapted to their diets. Every kind of animal has evolved a way for its body to get the nourishment it needs. Some animals only ever drink liquids, such as blood, or filter tiny particles from water, while others use muscles and jaws to tear solid food to pieces.

NILE CROCODILE
Strategy: Predator

Carnivores that must kill to obtain food not only need the skill to catch their prey, but also the strength to overpower it. Some predators rely on speed to chase prey down, but the Nile crocodile waits in ambush instead. It lurks submerged at a river's edge until a target comes to drink, then grabs the prey with its powerful jaws and pulls the struggling animal underwater to drown it.

Mighty bite
The crocodile's jaws can deliver a bite that has three times more force than a lion's.

Easy prey
Zebras are often attacked while crossing large rivers.

Many **predators**, such as spiders, **use disabling venom** to overpower their prey.

The **biggest living animal** – the blue whale – and the **biggest fish** – the whale shark – are both **filter feeders**.

151

LESSER FLAMINGO
Strategy: Filter feeder

The lesser flamingo is nourished almost entirely by the microscopic algae in African salt lakes. Each cupful of water from the lakes is a rich soup containing billions of algae, which the bird filters out with its unusual bill. By lowering its head upside down into the lake and pumping its tongue backwards and forwards like a piston, water gets drawn into and out of the long bill. A coating of minute brushes on the inner lining of the bill trap the algae, which are then rapidly swallowed by the hungry bird.

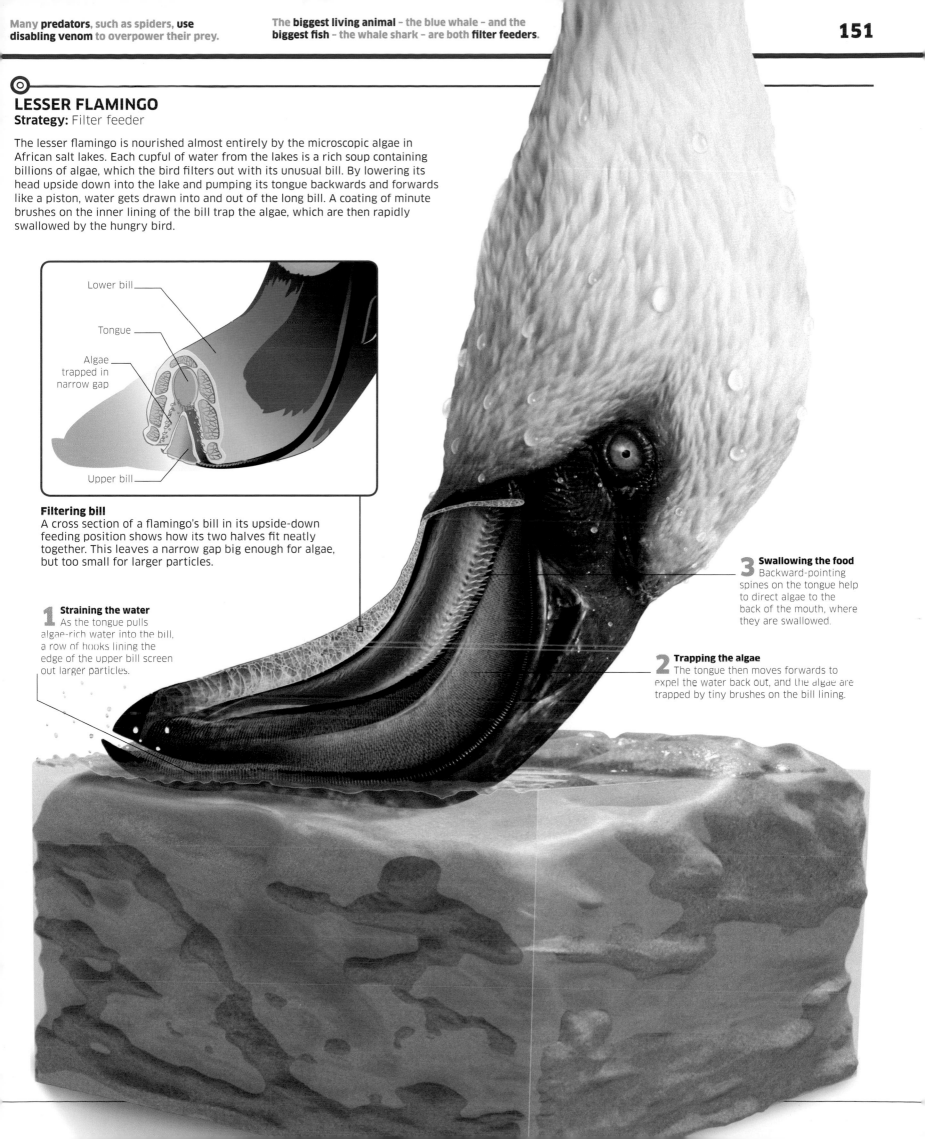

Lower bill

Tongue

Algae trapped in narrow gap

Upper bill

Filtering bill
A cross section of a flamingo's bill in its upside-down feeding position shows how its two halves fit neatly together. This leaves a narrow gap big enough for algae, but too small for larger particles.

1 Straining the water
As the tongue pulls algae-rich water into the bill, a row of hooks lining the edge of the upper bill screen out larger particles.

3 Swallowing the food
Backward-pointing spines on the tongue help to direct algae to the back of the mouth, where they are swallowed.

2 Trapping the algae
The tongue then moves forwards to expel the water back out, and the algae are trapped by tiny brushes on the bill lining.

Processing food

Eating is only part of the story of how the body gets nourishment. An animal's digestive system must then break down the food so that nutrients can reach cells.

Food contains vital ingredients called nutrients, such as sugars and vitamins. Most animals eat solid food, and the digestive system has to liquefy this food inside the body so these nutrients can seep into the bloodstream. Once dissolved in the blood, they are circulated around the body to get to where they are needed – inside cells.

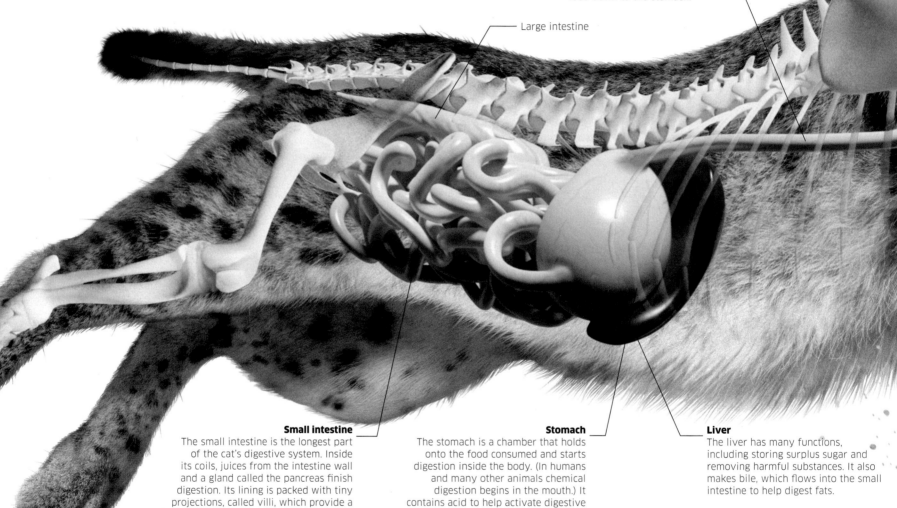

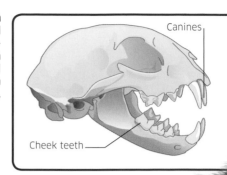

Carnivore teeth
Stabbing canines and sharp-edged, bone-crunching cheek teeth help the bobcat kill prey and bite through its skin and bones.

Canines

Cheek teeth

Oesophagus
The oesophagus (food) pipe carries lumps of swallowed food down to the stomach.

Large intestine

Small intestine
The small intestine is the longest part of the cat's digestive system. Inside its coils, juices from the intestine wall and a gland called the pancreas finish digestion. Its lining is packed with tiny projections, called villi, which provide a large surface area for absorbing nutrients.

Stomach
The stomach is a chamber that holds onto the food consumed and starts digestion inside the body. (In humans and many other animals chemical digestion begins in the mouth.) It contains acid to help activate digestive juices and to kill harmful microbes.

Liver
The liver has many functions, including storing surplus sugar and removing harmful substances. It also makes bile, which flows into the small intestine to help digest fats.

Releasing the nutrients

Biting and chewing by the mouth reduces food into manageable lumps for swallowing, but further processing is needed to extract the nutrients. Muscles in the wall of the digestive system churn food into a lumpy paste and mix it with digestive juices containing chemicals called enzymes. The enzymes help to drive chemical reactions that break big molecules into smaller ones, which are then absorbed into the blood.

Carbohydrates
Starch is digested into sugars, such as glucose.

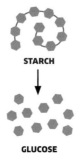

STARCH

GLUCOSE

Proteins
Proteins are digested into amino acids.

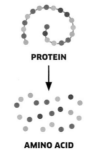

PROTEIN

AMINO ACID

Fats
Fats and oils are broken down to release fatty acids and glycerol.

FATS AND OILS

FATTY ACIDS

GLYCEROL

Digestive systems

A carnivorous bobcat and a herbivorous rabbit both have digestive systems filled with muscles and digestive juices to help break up their food. But they have important differences – each is adapted to the challenges of eating either chewy meat or tough vegetation.

There are more than
100 trillion bacteria
in the digestive tract.

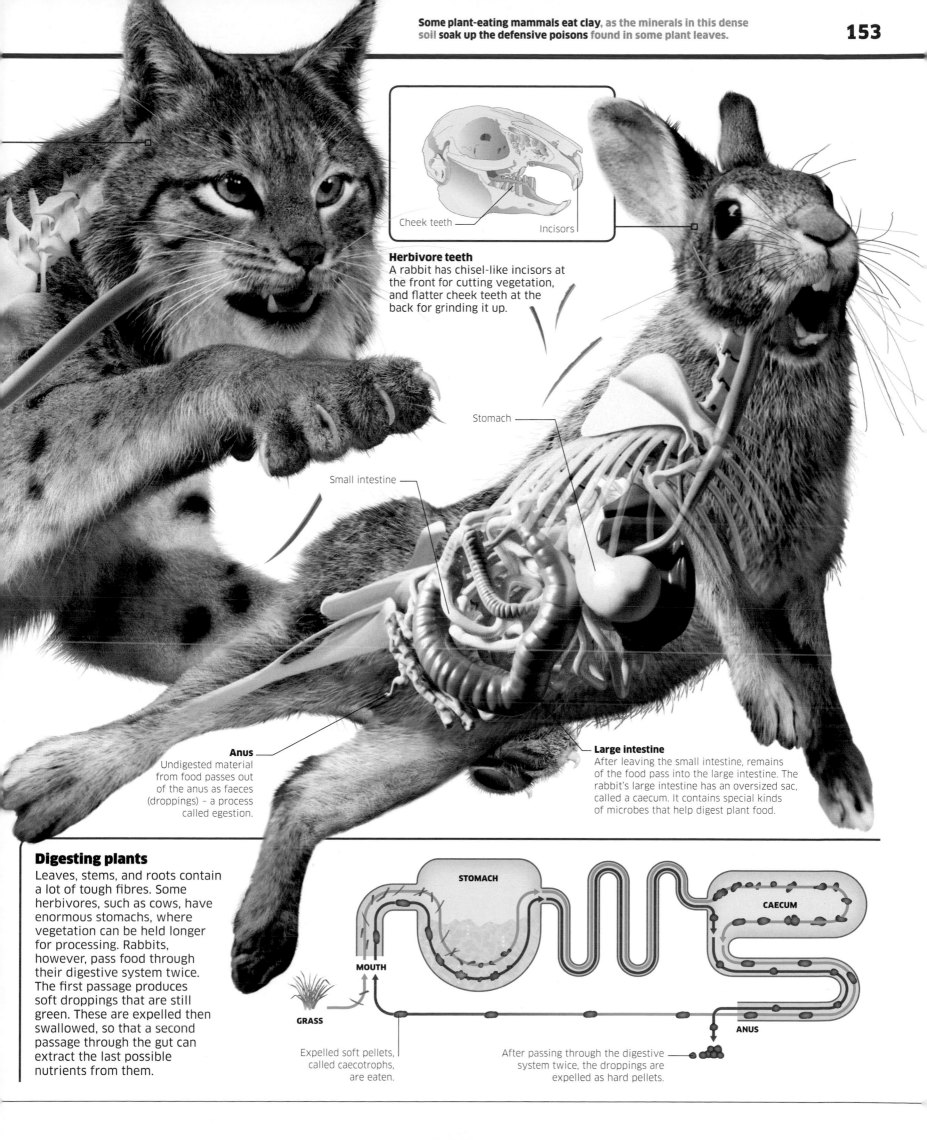

Some plant-eating mammals eat clay, as the minerals in this dense soil **soak up the defensive poisons** found in some plant leaves.

153

Cheek teeth

Incisors

Herbivore teeth
A rabbit has chisel-like incisors at the front for cutting vegetation, and flatter cheek teeth at the back for grinding it up.

Stomach

Small intestine

Anus
Undigested material from food passes out of the anus as faeces (droppings) – a process called egestion.

Large intestine
After leaving the small intestine, remains of the food pass into the large intestine. The rabbit's large intestine has an oversized sac, called a caecum. It contains special kinds of microbes that help digest plant food.

Digesting plants
Leaves, stems, and roots contain a lot of tough fibres. Some herbivores, such as cows, have enormous stomachs, where vegetation can be held longer for processing. Rabbits, however, pass food through their digestive system twice. The first passage produces soft droppings that are still green. These are expelled then swallowed, so that a second passage through the gut can extract the last possible nutrients from them.

STOMACH

CAECUM

MOUTH

GRASS

ANUS

Expelled soft pellets, called caecotrophs, are eaten.

After passing through the digestive system twice, the droppings are expelled as hard pellets.

Plant transport

To lift water to their topmost branches, trees need incredible transport systems. The tallest ones can pull water with the same force as a high-pressure hose.

Plants owe this remarkable ability to impressive engineering. Their trunks and stems are packed with bundles of microscopic pipes. Water and minerals are moved from the soil to the leaves, while food made in the leaves is sent around the entire plant.

3 Pull from above
Water evaporates from the moist tissues inside living leaves. The vapour it generates spills out into the surrounding atmosphere through pores called stomata. This water loss, known as transpiration, is replaced with water arriving from the ground in pipe-like xylem vessels.

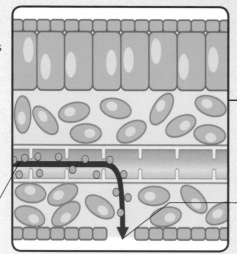

Water moves into the leaf in xylem vessels of the leaf's veins.

Water vapour escapes through pores called stomata.

2 Rising water
The microscopic xylem vessels carry unbroken columns of water through the stem all the way up to the leaves. Water molecules stick together, so as transpiration pulls water into the leaves, all the columns of water rise up through the stem – like water climbing through drinking straws. This is called the transpiration stream.

Transpiration

A tree's water transport system is incredibly efficient and, unlike the transport system of animals, does not require any energy from the organism. The sun's heat causes water to evaporate from the leaves, a process called transpiration, which triggers the tree to pull more water up from the ground.

Xylem vessels are made up of stacks of empty dead cells with holes in their ends.

Bark
Tough outer layers of bark serve to protect the tree's trunk from injury.

1 Absorption from below
Water seeps into the roots from the soil by a process called osmosis. It then passes into tubes called xylem vessels to join the transpiration stream upwards. Microscopic extensions to the root, called root hairs, help maximize the absorption area, so that the tree can pick up large amounts of water, and minerals too.

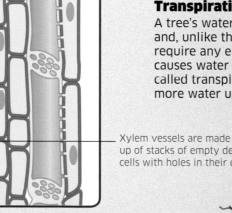
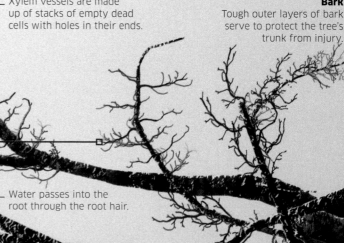

Water passes into the root through the root hair.

Xylem vessels

A mature oak tree can **transpire** more
than **a bathful of water** every day.

155

Phloem
The innermost layer of the
tree's bark, called the phloem,
transports food made by
photosynthesis in the leaves.

Cambium
A thin layer of
actively dividing
cells, the cambium
generates more xylem
and phloem as the
tree grows thicker.

Sapwood
This contains the xylem
vessels that stream
water up the tree.

Heartwood
This is made of old xylem
vessels that no longer carry
water, but help support the
weight of the tree.

Food distribution

Sugars and other food are made in the leaves
through photosynthesis (see pp.148–149). They
are then carried through pipes called phloem –
travelling to roots, flowers, and others parts
that cannot make food for themselves.

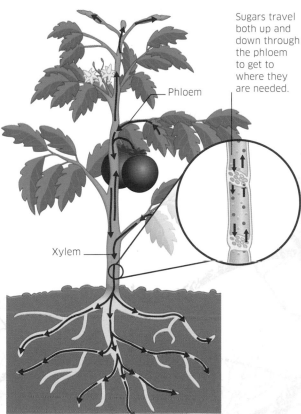

Sugars travel
both up and
down through
the phloem
to get to
where they
are needed.

Phloem

Xylem

Osmosis

When cell membranes stretch between two
solutions with different concentrations, water
automatically passes across to the higher
concentration by a process called osmosis.
This happens in plant roots – where root cell
membranes are situated between the weak
mineral solutions found in soil and the higher
concentrations inside the root cells.

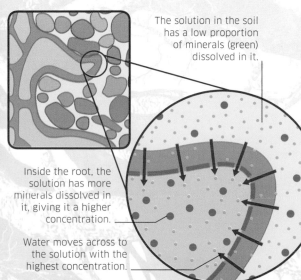

The solution in the soil
has a low proportion
of minerals (green)
dissolved in it.

Inside the root, the
solution has more
minerals dissolved in
it, giving it a higher
concentration.

Water moves across to
the solution with the
highest concentration.

Some invertebrates have blue blood – coloured by copper pigments.

70 times a minute is how fast the human heart beats on average.

The heart

The blue whale has the biggest heart of any animal: weighing in at 180 kg (400 lb) and standing as tall as a 12-year-old child. Containing four chambers, it is made of solid muscle, and contracts with a regular rhythm to pump blood out through the body's arteries. When its muscles relax, the pressure inside the chambers dips very low to pull in blood from the veins.

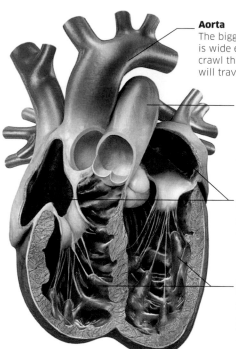

Aorta
The biggest artery in the blue whale is wide enough for a toddler to crawl through. Blood from here will travel around the body.

Pulmonary artery
Unlike in other arteries, the blood flowing through this artery does not carry oxygen, but travels to the lungs to pick it up.

Atria
The two small upper chambers of the heart are called atria. Atria pump blood into the ventricles.

Ventricles
The two larger chambers of the heart are called ventricles. The right ventricle pumps blood to the lungs, and the left pumps it around the rest of the body.

Network of vessels

Thousands of kilometres of blood vessels run through the body of a blue whale. Thick-walled arteries (shown in red) carry blood away from the heart and thin-walled veins (shown in blue) ferry it back. These both branch off countless times to form a network of microscopic capillaries (smaller blood vessels) that run between the cells.

Circulation

Blood is the essential life support system of many animals, transporting food and oxygen around the body and removing waste from cells.

Animals have trillions of cells that need support, and a vast network of tiny tubes called blood vessels stretches throughout their bodies in order to reach them all. A pumping heart keeps blood continually flowing through the blood vessels, and this bloodstream gathers food from the digestive system and oxygen from lungs or gills. When the blood reaches cells, these essentials pass inside, while waste moves back out of the cells and is then carried away by the blood to excretory organs, such as the kidneys.

10 tonnes of blood are contained within the body of a blue whale. Its heart **pumps two bathfuls with every beat.**

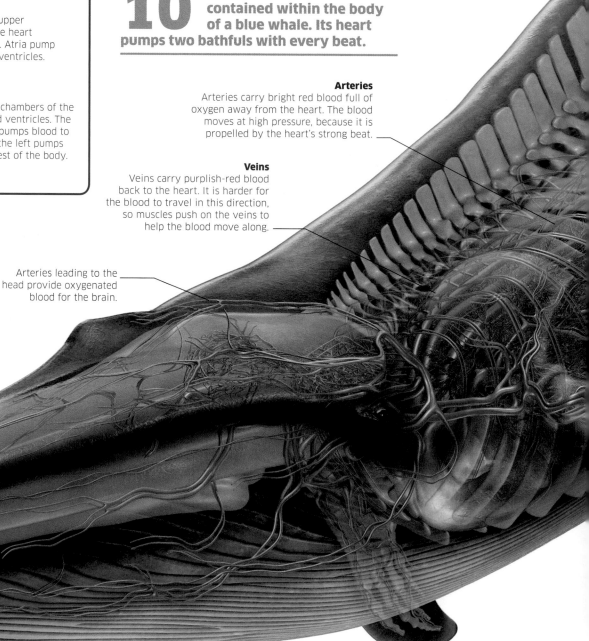

Arteries
Arteries carry bright red blood full of oxygen away from the heart. The blood moves at high pressure, because it is propelled by the heart's strong beat.

Veins
Veins carry purplish-red blood back to the heart. It is harder for the blood to travel in this direction, so muscles push on the veins to help the blood move along.

Arteries leading to the head provide oxygenated blood for the brain.

Tiny animals, such as shrews, have hearts
that can beat up to 1,000 times a minute.

It takes half a minute for blood to clot (thicken)
when exposed to air – helping to seal wounds.

157

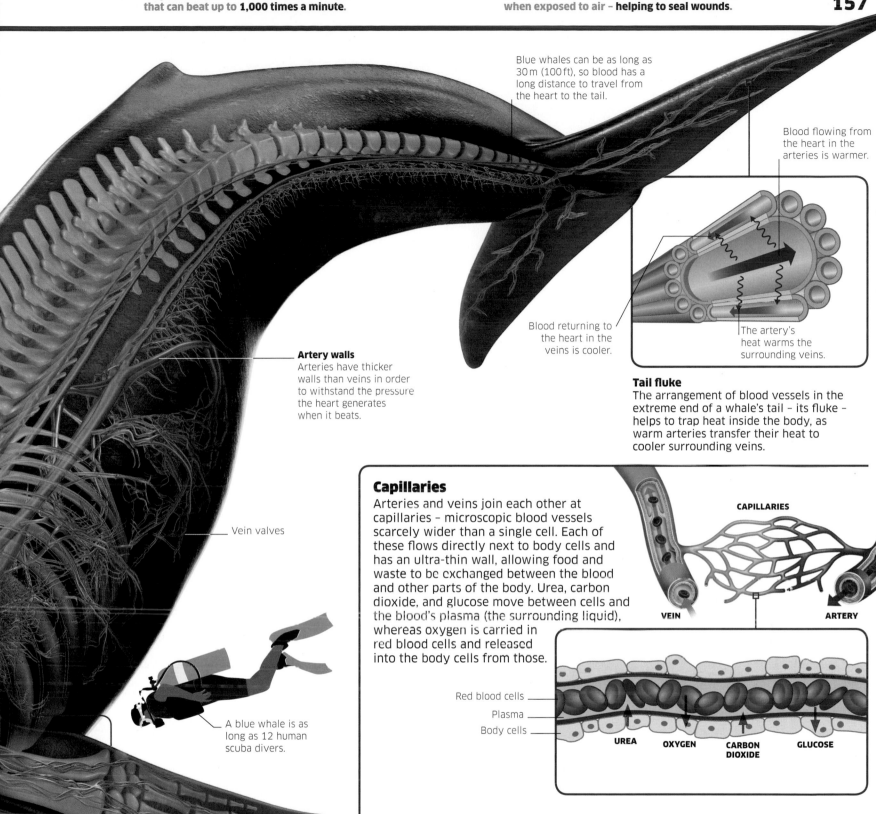

Blue whales can be as long as
30 m (100 ft), so blood has a
long distance to travel from
the heart to the tail.

Blood flowing from
the heart in the
arteries is warmer.

Artery walls
Arteries have thicker
walls than veins in order
to withstand the pressure
the heart generates
when it beats.

Blood returning to
the heart in the
veins is cooler.

The artery's
heat warms the
surrounding veins.

Tail fluke
The arrangement of blood vessels in the
extreme end of a whale's tail – its fluke –
helps to trap heat inside the body, as
warm arteries transfer their heat to
cooler surrounding veins.

Vein valves

Capillaries
Arteries and veins join each other at
capillaries – microscopic blood vessels
scarcely wider than a single cell. Each of
these flows directly next to body cells and
has an ultra-thin wall, allowing food and
waste to be exchanged between the blood
and other parts of the body. Urea, carbon
dioxide, and glucose move between cells and
the blood's plasma (the surrounding liquid),
whereas oxygen is carried in
red blood cells and released
into the body cells from those.

CAPILLARIES

VEIN

ARTERY

Red blood cells

Plasma

Body cells

UREA **OXYGEN** **CARBON DIOXIDE** **GLUCOSE**

A blue whale is as
long as 12 human
scuba divers.

Blood flows
through
the vein.

Double circulation
Mammals have a more efficient
circulation than fishes. Blood
pumped by a fish's heart moves
through the gills to pick up oxygen,
travels around the rest of the body,
and only then returns back to the
heart. However, in mammals, blood
returns to the heart directly after
the lungs. It then has more pressure
when it flows to the cells, making
exchanges easier. This is why
mammals have four chambers in
their hearts – both an upper and
lower chamber for each circuit.

GILL CAPILLARIES

LUNG CAPILLARIES

OTHER CAPILLARIES

OTHER CAPILLARIES

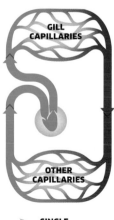

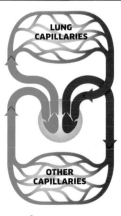

SINGLE CIRCULATION

DOUBLE CIRCULATION

Vein valves
One-way valves in
the veins close off
behind the blood as
it passes through, to
stop it flowing back
the other way.

Valves close behind
the blood, so it cannot
flow backwards.

Breathing with lungs

Land-living vertebrates, such as mammals, birds, and reptiles, breathe with lungs. These air-filled cavities sit inside the chest and have thin walls lined with blood vessels. When the animal breathes in and out, chest muscles expand and deflate the lungs, pulling in oxygen-rich air and removing waste carbon dioxide.

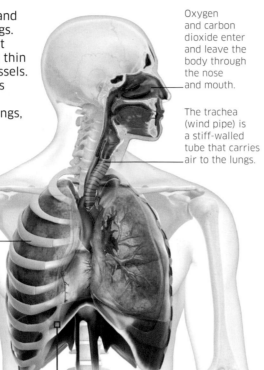

Oxygen and carbon dioxide enter and leave the body through the nose and mouth.

The trachea (wind pipe) is a stiff-walled tube that carries air to the lungs.

A sturdy rib cage protects the lungs, while chest muscles power them.

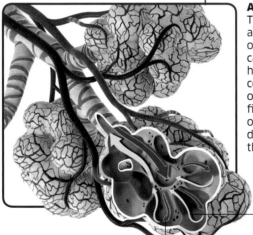

Alveoli

The lungs of mammals are made up of millions of microscopic sacs called alveoli. Each sac has an ultra-thin wall covered in a network of blood capillaries. This fine surface allows lots of oxygen and carbon dioxide to move between the air and the blood.

Waste carbon dioxide is transferred back from the blood to the lungs.

Oxygen moves through the very thin walls of the alveoli into the blood capillaries.

Diffusion

Oxygen and carbon dioxide are able to cross the microscopic membrane between the lungs and the blood by diffusion: a process by which molecules naturally move from an area where they are highly concentrated to one in which their numbers are fewer. This happens all around the body, as gases move between blood and respiring cells.

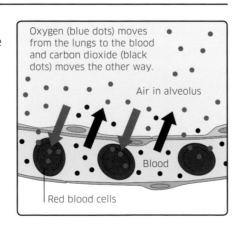

Oxygen (blue dots) moves from the lungs to the blood and carbon dioxide (black dots) moves the other way.

Air in alveolus

Blood

Red blood cells

Breathing

An animal breathes to supply its cells with oxygen, a vital resource that helps to burn up food and release much-needed energy around the body.

All organisms, including animals, plants, and microbes, get energy from respiration, a chemical reaction that happens inside cells. Most do this by reacting food with oxygen, producing carbon dioxide as a waste product. To drive the oxygen into the body, different animals have highly adapted respiratory systems, such as lungs or gills. These can exchange large quantities of gas, carrying oxygen to respiring cells in the bloodstream, and excreting waste carbon dioxide.

Oxygen travelling in the blood is attached to a pigment called haemoglobin –
the substance that gives blood its red colour.

Breathing with gills

Gills are feathery extensions of the body that splay out in water so that aquatic animals can breathe. The delicate, blood-filled gills of fish are protected inside chambers on either side of their mouth cavity. A fish breathes by opening its mouth to draw oxygen-rich water over its gills. Some fishes rely on the stream created as they swim forwards, but most use throat muscles to gulp water. Oxygen moves from the gills into the blood, while stale water emerges from the gill openings on either side of the head.

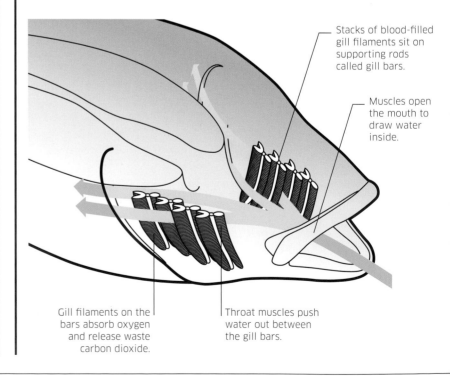

Stacks of blood-filled gill filaments sit on supporting rods called gill bars.

Muscles open the mouth to draw water inside.

Gill filaments on the bars absorb oxygen and release waste carbon dioxide.

Throat muscles push water out between the gill bars.

500 million alveoli are in the human lungs, providing an **enormous area for gas exchange**.

Some **underwater insects breathe with gills**, or even **carry bubbles of air** underwater with them.

159

Breathing with tracheae

Insects and related invertebrates have a breathing system that gets oxygen directly to their muscles. Instead of oxygen being carried in the blood, an intricate network of pipes reaches into the body from breathing holes called spiracles. Each pipe – known as a trachea – splits into tinier branches called tracheoles. The tracheoles are precisely arranged so that their tips penetrate the body's cells. This delivers oxygen-rich air from the surroundings deep into the insect, where respiration takes place.

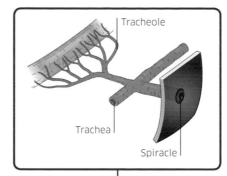

Tracheole

Trachea

Spiracle

Tracheoles

Microscopic air-filled tracheoles in the body of an insect perform a similar role to blood-filled capillaries in other animals: they pass oxygen into the cells, while carbon dioxide moves out. This direct and efficient system means cells get oxygen delivered straight to them.

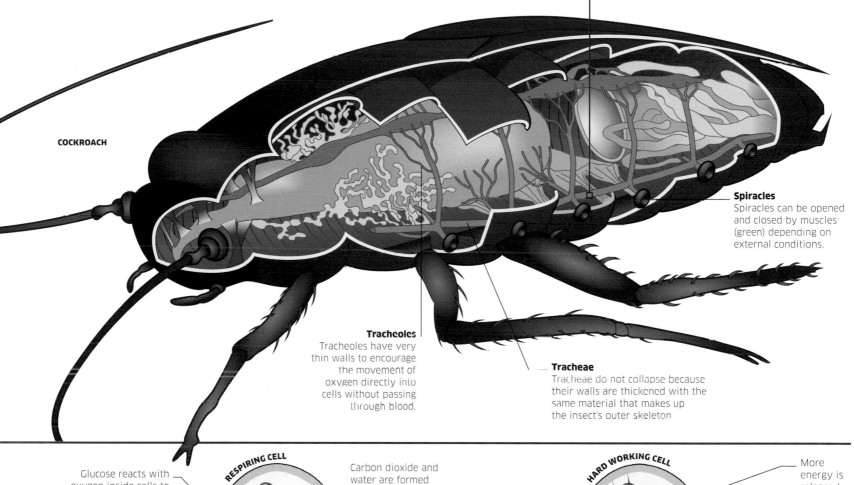

COCKROACH

Spiracles
Spiracles can be opened and closed by muscles (green) depending on external conditions.

Tracheoles
Tracheoles have very thin walls to encourage the movement of oxygen directly into cells without passing through blood.

Tracheae
Tracheae do not collapse because their walls are thickened with the same material that makes up the insect's outer skeleton.

RESPIRING CELL

Glucose reacts with oxygen inside cells to release energy.

Carbon dioxide and water are formed as waste products.

GLUCOSE

CARBON DIOXIDE

OXYGEN

WATER

HARD WORKING CELL

MORE GLUCOSE

MORE CARBON DIOXIDE

MORE OXYGEN

MORE WATER

More energy is released to power muscles when the horse is running.

CARBON DIOXIDE

OXYGEN

CARBON DIOXIDE

OXYGEN

Cellular respiration

The oxygen an animal brings into its body is used in a chemical process called respiration. This mainly takes place in capsules within cells, called mitochondria. Here, energy-rich foods, such as sugars (glucose), are broken down into smaller molecules to release usable energy. The more active an animal is, the more oxygen it needs to keep this reaction running. While oxygen is crucial for most respiration, animals under physical pressure can release a tiny bit of extra energy without oxygen – a process known as anaerobic respiration.

Glucose from digested food is stored inside cells until it is needed.

An active animal breathes faster and deeper to supply its cells with more oxygen to get more energy.

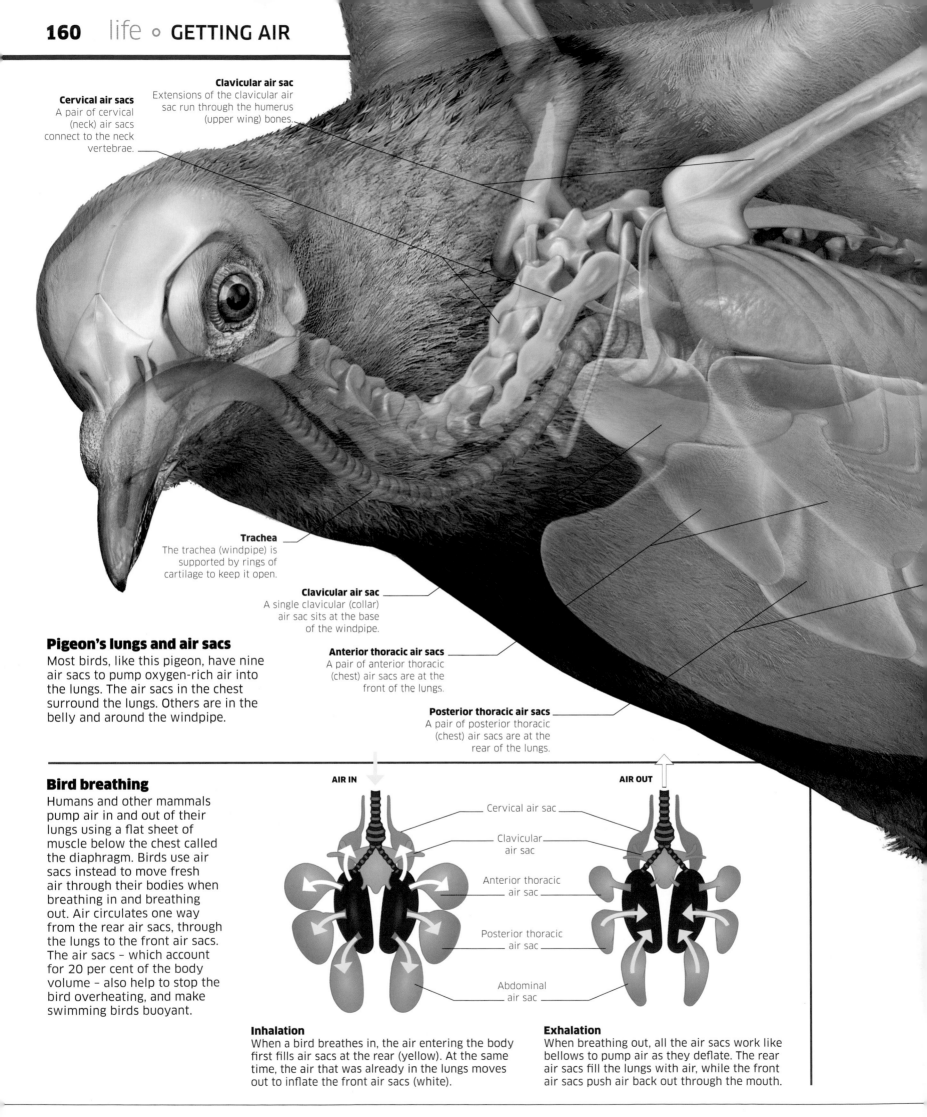

Cervical air sacs
A pair of cervical (neck) air sacs connect to the neck vertebrae.

Clavicular air sac
Extensions of the clavicular air sac run through the humerus (upper wing) bones.

Trachea
The trachea (windpipe) is supported by rings of cartilage to keep it open.

Clavicular air sac
A single clavicular (collar) air sac sits at the base of the windpipe.

Anterior thoracic air sacs
A pair of anterior thoracic (chest) air sacs are at the front of the lungs.

Posterior thoracic air sacs
A pair of posterior thoracic (chest) air sacs are at the rear of the lungs.

Pigeon's lungs and air sacs

Most birds, like this pigeon, have nine air sacs to pump oxygen-rich air into the lungs. The air sacs in the chest surround the lungs. Others are in the belly and around the windpipe.

Bird breathing

Humans and other mammals pump air in and out of their lungs using a flat sheet of muscle below the chest called the diaphragm. Birds use air sacs instead to move fresh air through their bodies when breathing in and breathing out. Air circulates one way from the rear air sacs, through the lungs to the front air sacs. The air sacs – which account for 20 per cent of the body volume – also help to stop the bird overheating, and make swimming birds buoyant.

AIR IN

AIR OUT

Cervical air sac

Clavicular air sac

Anterior thoracic air sac

Posterior thoracic air sac

Abdominal air sac

Inhalation
When a bird breathes in, the air entering the body first fills air sacs at the rear (yellow). At the same time, the air that was already in the lungs moves out to inflate the front air sacs (white).

Exhalation
When breathing out, all the air sacs work like bellows to pump air as they deflate. The rear air sacs fill the lungs with air, while the front air sacs push air back out through the mouth.

Getting air

Birds need plenty of energy to fuel active lives. Their beautifully efficient system for getting oxygen to cells is unique – no other animal has one quite like it.

The key to a bird's breathing system lies in big, air-filled sacs that pack the body. They help supply air to the lungs, which, unlike human lungs, are small and rigid. Breathing makes the sacs inflate and deflate like balloons, sweeping fresh air through the lungs. Oxygen continually seeps into the blood and circulates to the cells, so they can release energy in respiration.

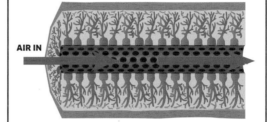

Parabronchi (air-filled tubes) bring air to the air vessels.

AIR IN

Microscopic tubes called capillaries carry blood.

Gaseous exchange
Each lung is filled with tiny air-filled vessels intermingled with microscopic blood vessels, helping to bring oxygen as close to the blood as possible.

Abdominal air sacs
The biggest pair of air sacs are in the abdomen (belly) of the body.

Breathing at high altitudes
Travelling at high altitudes poses a special problem for some high-flying birds, as the air gets so thin that there is little oxygen. The migration route of bar-headed geese takes them over the Himalayas – the highest any known bird has flown. To cope with this, they have bigger lungs than other waterfowl and can breathe more deeply, while the pigment in their blood (haemoglobin) traps oxygen in the thin air especially well.

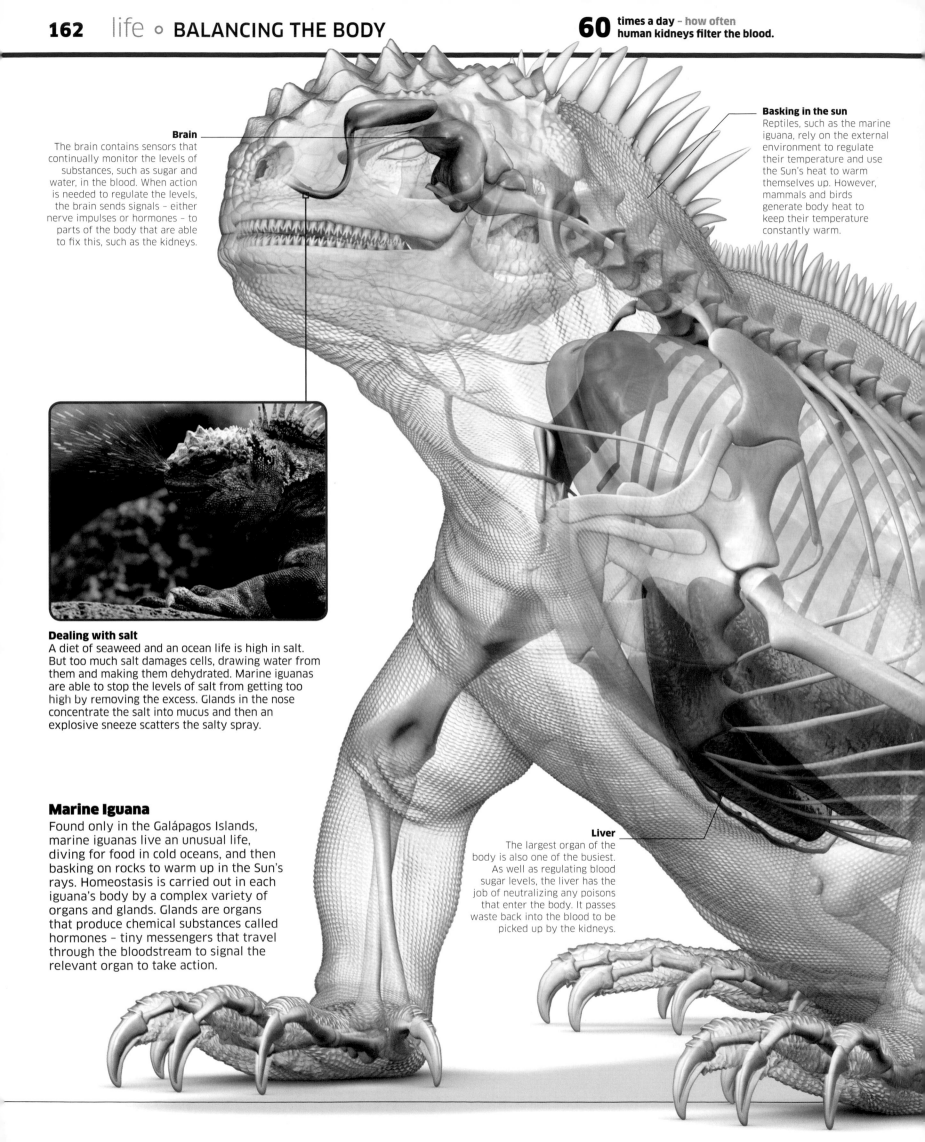

162 life ∘ BALANCING THE BODY

60 times a day – how often human kidneys filter the blood.

Brain
The brain contains sensors that continually monitor the levels of substances, such as sugar and water, in the blood. When action is needed to regulate the levels, the brain sends signals – either nerve impulses or hormones – to parts of the body that are able to fix this, such as the kidneys.

Basking in the sun
Reptiles, such as the marine iguana, rely on the external environment to regulate their temperature and use the Sun's heat to warm themselves up. However, mammals and birds generate body heat to keep their temperature constantly warm.

Dealing with salt
A diet of seaweed and an ocean life is high in salt. But too much salt damages cells, drawing water from them and making them dehydrated. Marine iguanas are able to stop the levels of salt from getting too high by removing the excess. Glands in the nose concentrate the salt into mucus and then an explosive sneeze scatters the salty spray.

Marine Iguana
Found only in the Galápagos Islands, marine iguanas live an unusual life, diving for food in cold oceans, and then basking on rocks to warm up in the Sun's rays. Homeostasis is carried out in each iguana's body by a complex variety of organs and glands. Glands are organs that produce chemical substances called hormones – tiny messengers that travel through the bloodstream to signal the relevant organ to take action.

Liver
The largest organ of the body is also one of the busiest. As well as regulating blood sugar levels, the liver has the job of neutralizing any poisons that enter the body. It passes waste back into the blood to be picked up by the kidneys.

37 °C (99°F) – the ideal temperature for enzymes in the human body to function.

500 chemical reactions are performed by the human liver cells in order to carry out its everyday functions.

163

Balancing the body

While external conditions may change from rain to shine, the internal environment of an animal's body is carefully controlled to ensure the vital processes of life can take place.

This balancing act is called homeostasis. Complex backboned animals have especially good systems of homeostasis that regulate factors such as body temperature, blood sugar levels, and water levels. Alongside this, other areas of the body carry out a process called excretion to remove waste, which can be harmful if left to accumulate. This continual regulation gives the body the right set of conditions to carry out all the functions of life, such as processing food and releasing energy.

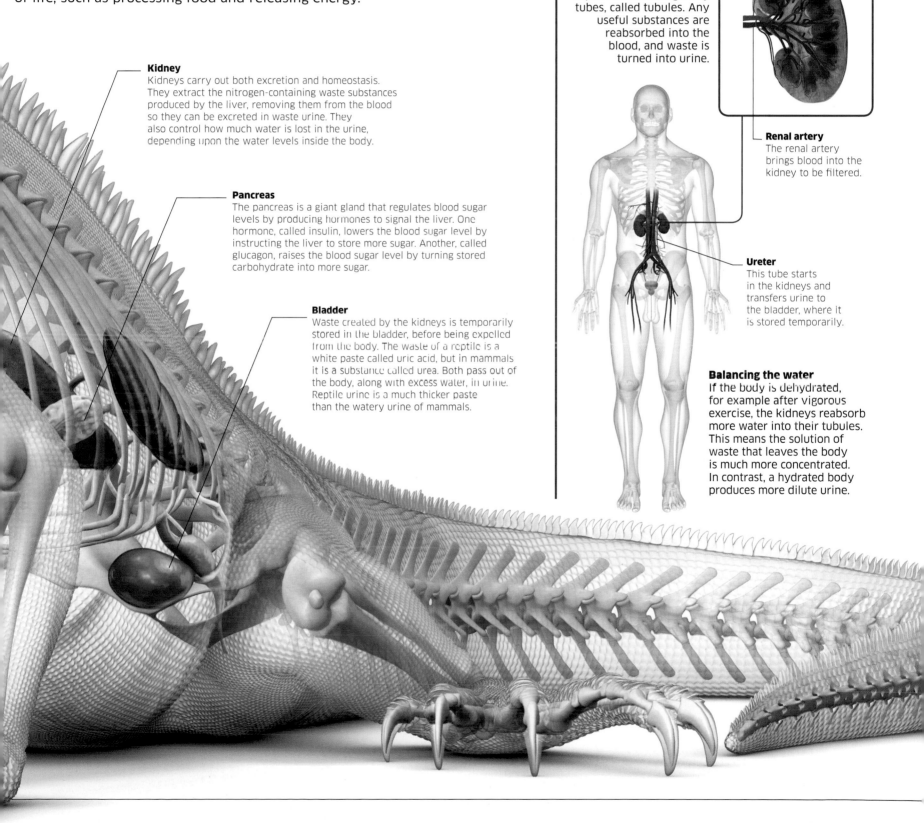

Kidney

Kidneys carry out both excretion and homeostasis. They extract the nitrogen-containing waste substances produced by the liver, removing them from the blood so they can be excreted in waste urine. They also control how much water is lost in the urine, depending upon the water levels inside the body.

Pancreas

The pancreas is a giant gland that regulates blood sugar levels by producing hormones to signal the liver. One hormone, called insulin, lowers the blood sugar level by instructing the liver to store more sugar. Another, called glucagon, raises the blood sugar level by turning stored carbohydrate into more sugar.

Bladder

Waste created by the kidneys is temporarily stored in the bladder, before being expelled from the body. The waste of a reptile is a white paste called uric acid, but in mammals it is a substance called urea. Both pass out of the body, along with excess water, in urine. Reptile urine is a much thicker paste than the watery urine of mammals.

Urinary system

Although mammal and reptile kidneys differ in their shape, both contain a complex system of blood vessels and tubes to filter the blood of waste products. These, along with the bladder, make up the urinary system. As well as removing excess water, most kidneys can also excrete in urine any unwanted salt, unlike those of the marine iguana.

Filtering the blood

Kidneys filter liquid, containing waste, directly from the blood. This liquid drains through tiny tubes, called tubules. Any useful substances are reabsorbed into the blood, and waste is turned into urine.

Renal artery

The renal artery brings blood into the kidney to be filtered.

Ureter

This tube starts in the kidneys and transfers urine to the bladder, where it is stored temporarily.

Balancing the water

If the body is dehydrated, for example after vigorous exercise, the kidneys reabsorb more water into their tubules. This means the solution of waste that leaves the body is much more concentrated. In contrast, a hydrated body produces more dilute urine.

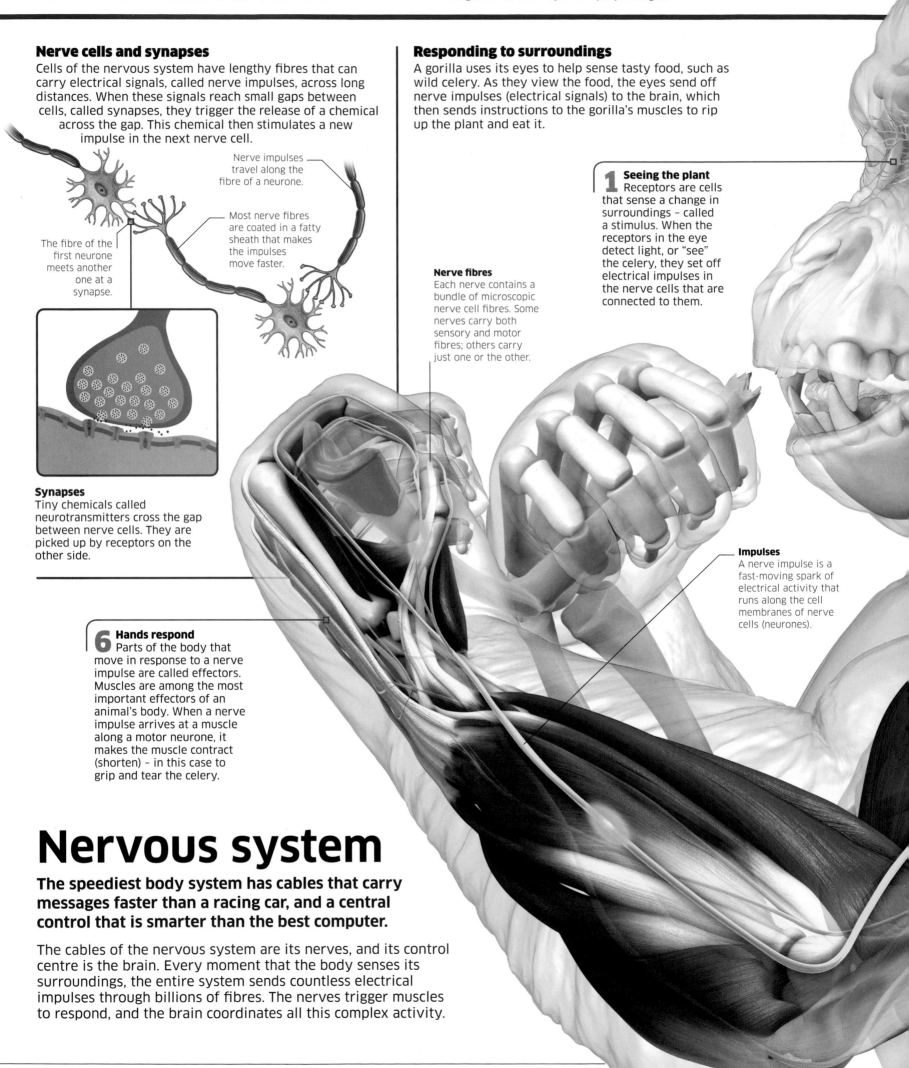

Nerve cells and synapses

Cells of the nervous system have lengthy fibres that can carry electrical signals, called nerve impulses, across long distances. When these signals reach small gaps between cells, called synapses, they trigger the release of a chemical across the gap. This chemical then stimulates a new impulse in the next nerve cell.

Nerve impulses travel along the fibre of a neurone.

Most nerve fibres are coated in a fatty sheath that makes the impulses move faster.

The fibre of the first neurone meets another one at a synapse.

Synapses
Tiny chemicals called neurotransmitters cross the gap between nerve cells. They are picked up by receptors on the other side.

Responding to surroundings

A gorilla uses its eyes to help sense tasty food, such as wild celery. As they view the food, the eyes send off nerve impulses (electrical signals) to the brain, which then sends instructions to the gorilla's muscles to rip up the plant and eat it.

1 Seeing the plant
Receptors are cells that sense a change in surroundings – called a stimulus. When the receptors in the eye detect light, or "see" the celery, they set off electrical impulses in the nerve cells that are connected to them.

Nerve fibres
Each nerve contains a bundle of microscopic nerve cell fibres. Some nerves carry both sensory and motor fibres; others carry just one or the other.

Impulses
A nerve impulse is a fast-moving spark of electrical activity that runs along the cell membranes of nerve cells (neurones).

6 Hands respond
Parts of the body that move in response to a nerve impulse are called effectors. Muscles are among the most important effectors of an animal's body. When a nerve impulse arrives at a muscle along a motor neurone, it makes the muscle contract (shorten) – in this case to grip and tear the celery.

Nervous system

The speediest body system has cables that carry messages faster than a racing car, and a central control that is smarter than the best computer.

The cables of the nervous system are its nerves, and its control centre is the brain. Every moment that the body senses its surroundings, the entire system sends countless electrical impulses through billions of fibres. The nerves trigger muscles to respond, and the brain coordinates all this complex activity.

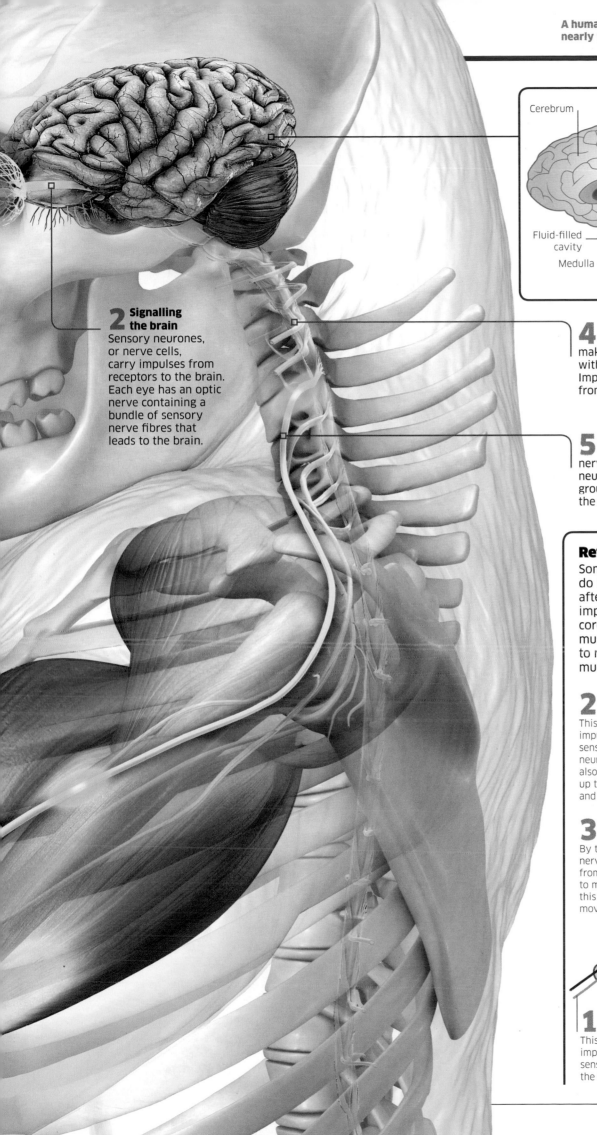

CROSS SECTION

Cerebrum

Fluid-filled
cavity

Medulla

Cerebellum

**3 Coordinating
a response**
The brain coordinates
where impulses go in
order to control the
body's behaviour. The
cerebrum manages
complex actions that
demand intelligence, like
peeling and breaking
up food. More routine
actions, such as walking,
are controlled by the
cerebellum, while the
medulla effects internal
functions, like breathing.

**2 Signalling
the brain**
Sensory neurones,
or nerve cells,
carry impulses from
receptors to the brain.
Each eye has an optic
nerve containing a
bundle of sensory
nerve fibres that
leads to the brain.

4 Travelling onwards
Together with the brain, the spinal cord
makes up the central nervous system. It works
with the brain to pass signals around the body.
Impulses travelling from the brain branch off
from the spinal cord to motor neurones.

5 Signalling the muscles
Cells that carry impulses from the central
nervous system to muscles are called motor
neurones. Bundles of motor neurone fibres are
grouped into nerves that run all the way from
the spinal cord to the limb muscles.

Reflex actions

Some automatic responses, called reflex actions,
do not involve the brain, such as when you recoil
after touching something hot. In these instances,
impulses travel from the sense organs to the spinal
cord, where relay neurones pass the signal to the
muscles. Bypassing the brain allows the impulses
to reach the effectors and generate a response
much more quickly.

**2 Relay
neurone**
This passes nerve
impulses from
sensory to motor
neurones. It can
also pass signals
up the spinal cord
and to the brain.

**CROSS SECTION OF
SPINAL CORD**

**3 Motor
neurone**
By transmitting
nerve impulses
from spinal cord
to muscles,
this triggers
movement.

**SPINE
VERTEBRA**

**1 Sensory
neurone**
This carries nerve
impulses from a
sense organ into
the spinal cord.

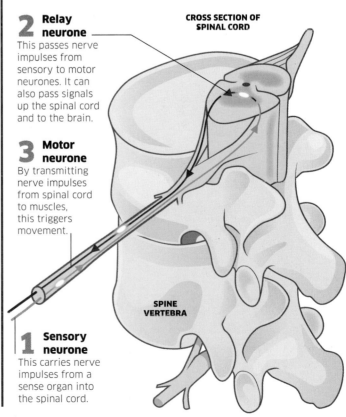

Vision

Eyes packed with light-sensitive cells enable animals to see. Vertebrates, such as humans, have two camera-like eyes that focus light onto the back of the eye. But some invertebrates rely on many more eyes: the giant clam has hundreds of tiny eyes scattered over its body. Each animal's eyes are specialised in different ways. Some are so sensitive that they can pick up the faintest light in the dark of night, or in the deep sea.

Four-eyed fish
When it swims at the surface, this fish's split-level eyes help it to focus on objects above and below the water.

Long-legged fly
Flies and many other insects have compound eyes – made up of thousands of tiny lenses.

Tarsier
This primate's eyes are the biggest of any mammal, when compared to the size of its head. They help it see well at night.

Touch

Animals have receptors in their skin that sense when other things come into contact with their body. Some receptors only pick up firm pressures, while others are sensitive to the lightest of touches. Receptor cells are especially concentrated in parts of the body that rely a lot on feeling textures or movement. Human fingertips are crammed with touch receptors, as are the whiskers of many cats, and the unusual nose of the star-nosed mole.

Star-nosed mole
The fleshy nose tentacles of this animal have six times more touch receptors than a human hand.

Senses

Animals sense their surroundings using organs that are triggered by light, sound, chemicals, or a whole range of other cues.

Sense organs are part of an animal's nervous system. They contain special cells called receptors that are stimulated by changes in the environment, and pass on signals to the brain and the rest of the body. Through these organs animals can gain a wealth of information about their surroundings, equipping them to react to threats or opportunities. Each kind of animal has sense organs that are best suited to the way it lives.

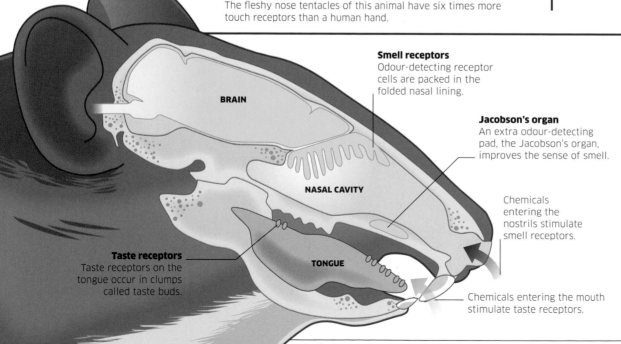

Smell receptors
Odour-detecting receptor cells are packed in the folded nasal lining.

BRAIN

Jacobson's organ
An extra odour-detecting pad, the Jacobson's organ, improves the sense of smell.

NASAL CAVITY

Chemicals entering the nostrils stimulate smell receptors.

Taste receptors
Taste receptors on the tongue occur in clumps called taste buds.

TONGUE

Chemicals entering the mouth stimulate taste receptors.

Taste and smell

Smelling and tasting are two very similar senses, as they both detect chemicals. The tongue has receptors that taste the chemicals dissolved in food and drink, and receptors inside the nose cavities pick up the chemicals in odours. Some animals that are especially reliant on chemical senses, or that do not have receptors elsewhere, have a concentrated patch of receptors in the roof of their mouth, called a Jacobson's organ.

Mouse senses

Like most mammals, a mouse has a keen nose. It uses smell to communicate with others of its kind: signalling a territorial claim or a willingness to mate. A mouse's tongue detects tastes in food, and both tongue and nose send signals to the brain.

Sharks have **sensory pits on their snouts** for detecting the **electrical fields** of prey.

Blood-sucking **mosquitoes are attracted to the body odour and carbon dioxide** produced by their victims.

167

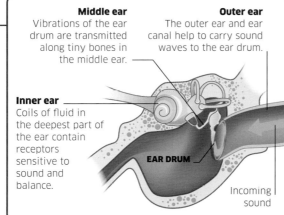

Pinna

Hearing

Animals hear because their ears contain receptors that are sensitive to sound waves. As the waves enter the ear, they vibrate a membrane called an ear drum. The vibrations pass along a chain of tiny bones until they reach the receptors within the inner ear.

Middle ear
Vibrations of the ear drum are transmitted along tiny bones in the middle ear.

Outer ear
The outer ear and ear canal help to carry sound waves to the ear drum.

Inner ear
Coils of fluid in the deepest part of the ear contain receptors sensitive to sound and balance.

EAR DRUM

Incoming sound

Bat-eared fox
In mammals, each ear opening is surrounded by a fleshy funnel for collecting sound, called a pinna. The desert-living bat-eared fox has such large pinnae that it also uses them to radiate warmth to stop it from overheating.

Hearing ranges
The pitch, or frequency, of a sound is measured in hertz (Hz): the number of vibrations per second. Different kinds of animals detect different ranges of pitch, and many are sensitive to ultrasound and infrasound that are beyond the hearing of humans.

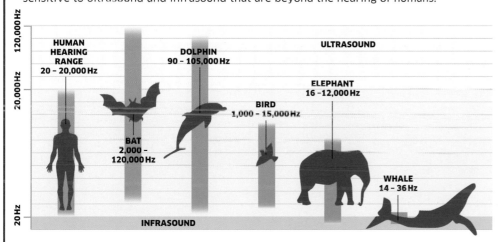

120,000 Hz
20,000 Hz
20 Hz

HUMAN HEARING RANGE
20 – 20,000 Hz

ULTRASOUND

DOLPHIN
90 – 105,000 Hz

BAT
2,000 – 120,000 Hz

BIRD
1,000 – 15,000 Hz

ELEPHANT
16 –12,000 Hz

WHALE
14 – 36 Hz

INFRASOUND

Other ways of sensing the world

The lives of many kinds of animals rely on quite extraordinary sensory systems. Some have peculiar types of receptors that are not found in other animals. These give them the power to sense their surroundings in ways that seem quite unfamiliar to us – such as by picking up electrical or magnetic fields.

Electroreception
The rubbery bill of the platypus – an aquatic egg-laying mammal – contains receptors that detect electrical signals coming from the muscles of moving prey. They help the platypus find worms and crayfish in murky river waters.

Echolocation
Bats and dolphins use echolocation to navigate and find food. By calling out and listening for the echoes bouncing back from nearby objects, they can work out the positions of obstacles and prey.

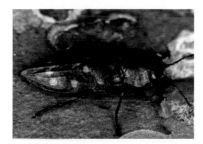

Fire detection
Most animals flee from fire, but the fire beetle thrives near flames. Its receptors pick up the infrared radiation coming from a blaze, drawing it to burnt-out trees where it can breed undisturbed by predators.

Magnetoreception
Birds can sense the earth's magnetic field. By combining this with information about the time of day and position of the Sun or stars, they can navigate their way on long-distance migrations.

Balance
All backboned animals have balance receptors in their ears to sense the position of their head and tell up from down. These help humans walk upright and stop climbing animals, such as capuchin monkeys, from falling out of trees.

Time
Tiny animals, such as insects, experience time more slowly because their senses can process more information every second. Compared with humans, houseflies see everything in slow motion – helping them to dodge predators.

Snake senses
A snake's tongue has no receptors and instead is used for transferring odours and tastes from prey and enemies to its sense organ in the roof of the mouth. A small nostril picks up additional smells.

Jacobson's organ
Chemicals on the tongue tip are transferred here to be detected.

Smell receptors pass signals along nerves to the brain.

NOSTRIL

BRAIN

Chemicals from the air, surfaces, or food are picked up by the tongue.

The forked tip helps to collect chemicals coming from both the left and the right.

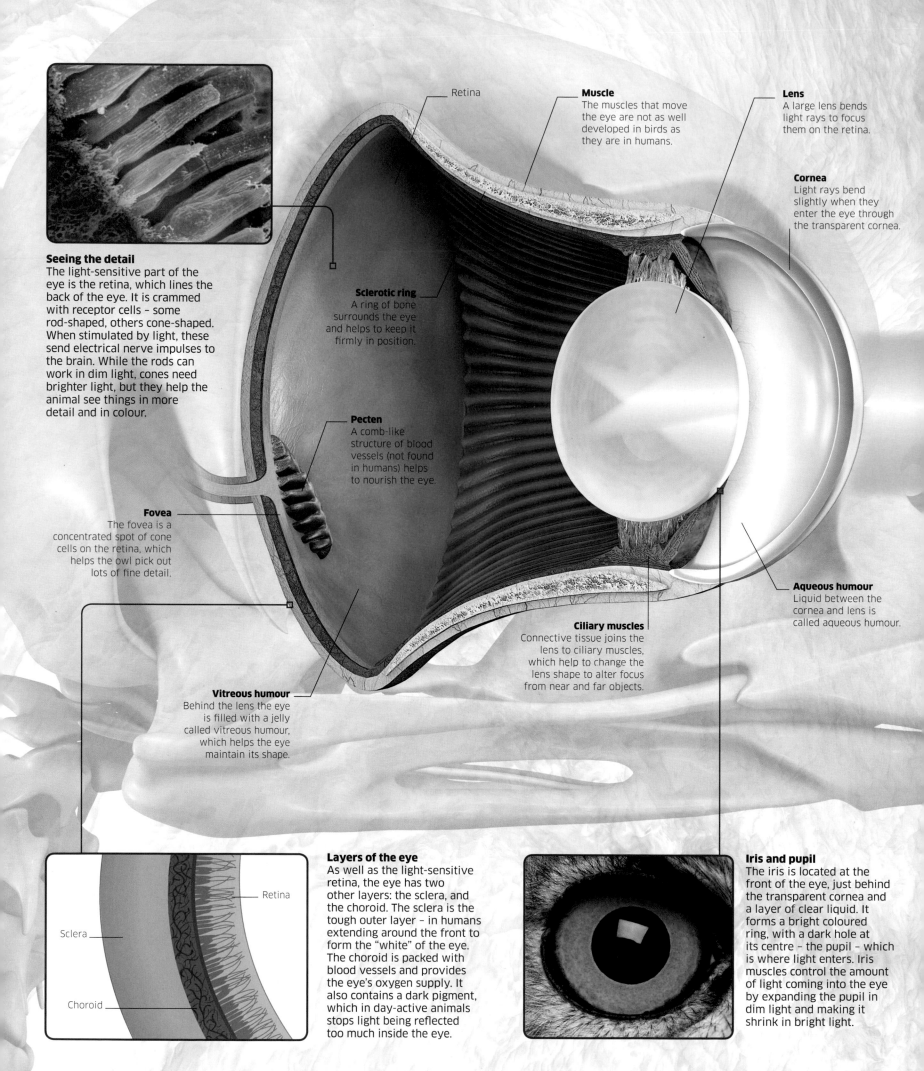

Retina

Muscle
The muscles that move the eye are not as well developed in birds as they are in humans.

Lens
A large lens bends light rays to focus them on the retina.

Cornea
Light rays bend slightly when they enter the eye through the transparent cornea.

Seeing the detail
The light-sensitive part of the eye is the retina, which lines the back of the eye. It is crammed with receptor cells – some rod-shaped, others cone-shaped. When stimulated by light, these send electrical nerve impulses to the brain. While the rods can work in dim light, cones need brighter light, but they help the animal see things in more detail and in colour.

Sclerotic ring
A ring of bone surrounds the eye and helps to keep it firmly in position.

Pecten
A comb-like structure of blood vessels (not found in humans) helps to nourish the eye.

Fovea
The fovea is a concentrated spot of cone cells on the retina, which helps the owl pick out lots of fine detail.

Aqueous humour
Liquid between the cornea and lens is called aqueous humour.

Ciliary muscles
Connective tissue joins the lens to ciliary muscles, which help to change the lens shape to alter focus from near and far objects.

Vitreous humour
Behind the lens the eye is filled with a jelly called vitreous humour, which helps the eye maintain its shape.

Retina

Sclera

Choroid

Layers of the eye
As well as the light-sensitive retina, the eye has two other layers: the sclera, and the choroid. The sclera is the tough outer layer – in humans extending around the front to form the "white" of the eye. The choroid is packed with blood vessels and provides the eye's oxygen supply. It also contains a dark pigment, which in day-active animals stops light being reflected too much inside the eye.

Iris and pupil
The iris is located at the front of the eye, just behind the transparent cornea and a layer of clear liquid. It forms a bright coloured ring, with a dark hole at its centre – the pupil – which is where light enters. Iris muscles control the amount of light coming into the eye by expanding the pupil in dim light and making it shrink in bright light.

Nocturnal mammals, such as cats, have a light-reflective layer behind their retinas, which makes their eyes shine when illuminated.

No animal can see in pitch darkness. All animals must detect at least a small amount of light to have vision.

169

Vision

The ability to see allows all animals to build up a detailed picture of their surroundings, vital for finding food and avoiding danger.

When an animal sees the world, its eyes pick up light and use lenses to focus this onto receptor cells that are light-sensitive. These cells then send signals to the brain, which composes a visual image of everything in the field of view. For animals with the best vision, the image can be finely detailed – even when the light is poor.

Night eyes

The eyes of birds are so big in proportion to their head that they are largely fixed inside their sockets. This means a bird must rotate its flexible neck to look around. Owl eyes, like those of many nocturnal birds, are especially large and are designed for good night vision. Their unusual shape creates room for a larger space at the back of the eye, packed with extra light-sensitive cells.

Seeing colour

Receptor cells called cones are what allow animals to see colour. These detect different light wavelengths – from short blue wavelengths to long red ones. Animals with more types of cones can see more colours, but those with just one are only able to see the world in black and white.

Humans have three kinds of cones, but dolphins have only one.

Three cones help humans see three primary colours: red, green, and blue, plus all their combinations.

Many day-flying birds have one more type of cone than humans, which allows them to see ultraviolet.

Near and far

The eye's lens focuses light onto the retina, and can change shape to better focus on either closer objects, or those further away. A ring of muscle controls this shape. It contracts to make the lens rounder for near focus, and relaxes to pull the lens flatter for distant focus.

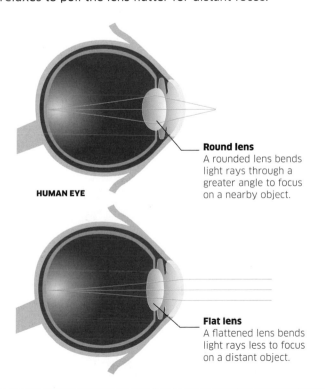

HUMAN EYE

Round lens
A rounded lens bends light rays through a greater angle to focus on a nearby object.

Flat lens
A flattened lens bends light rays less to focus on a distant object.

Binocular vision

When two forward-facing eyes have overlapping fields of view, this is called binocular vision. This gives an animal a three-dimensional view of the world, helping it to judge distance – a skill especially important for predators that hunt prey. Other animals with eyes on the sides of their head have a narrower range of binocular vision, but better all-round vision.

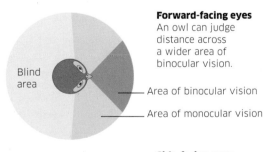

Forward-facing eyes
An owl can judge distance across a wider area of binocular vision.

Blind area

Area of binocular vision

Area of monocular vision

Side-facing eyes
Most other birds, such as snipes, only have binocular vision in front and behind, but can see through a wider total area.

Monocular vision

Blind area

Binocular vision

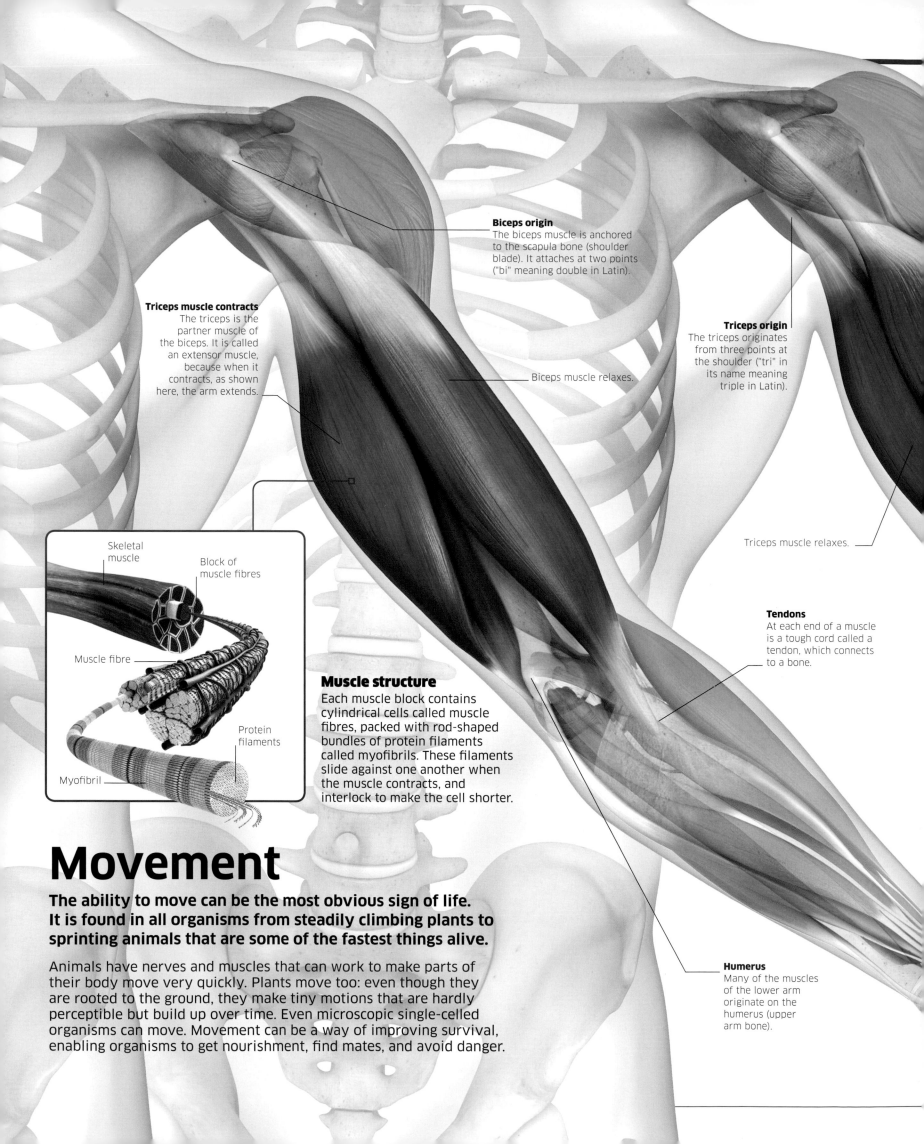

Biceps origin
The biceps muscle is anchored to the scapula bone (shoulder blade). It attaches at two points ("bi" meaning double in Latin).

Triceps muscle contracts
The triceps is the partner muscle of the biceps. It is called an extensor muscle, because when it contracts, as shown here, the arm extends.

Biceps muscle relaxes.

Triceps origin
The triceps originates from three points at the shoulder ("tri" in its name meaning triple in Latin).

Triceps muscle relaxes.

Skeletal muscle

Block of muscle fibres

Muscle fibre

Protein filaments

Myofibril

Muscle structure
Each muscle block contains cylindrical cells called muscle fibres, packed with rod-shaped bundles of protein filaments called myofibrils. These filaments slide against one another when the muscle contracts, and interlock to make the cell shorter.

Tendons
At each end of a muscle is a tough cord called a tendon, which connects to a bone.

Movement

The ability to move can be the most obvious sign of life. It is found in all organisms from steadily climbing plants to sprinting animals that are some of the fastest things alive.

Animals have nerves and muscles that can work to make parts of their body move very quickly. Plants move too: even though they are rooted to the ground, they make tiny motions that are hardly perceptible but build up over time. Even microscopic single-celled organisms can move. Movement can be a way of improving survival, enabling organisms to get nourishment, find mates, and avoid danger.

Humerus
Many of the muscles of the lower arm originate on the humerus (upper arm bone).

The snapping movement of a **Venus flytrap** is controlled by an **electrical impulse**.

Heart muscle is **the only muscle that can spontaneously contract** without being triggered by a nerve impulse.

171

It takes a muscle around
twice as long
to relax than to contract.

Biceps muscle contracts
The biceps of the upper arm is called a flexor muscle, because when it contracts, as shown here, it pulls on the lower arm to flex (bend) the elbow joint.

Finger movement
There are no muscles in the fingers – only tendons. These connect to the muscles in the rest of the hand.

Forearm muscles
The muscles in the lower arm control the complex movements of the wrist, hand, and fingers.

Working in pairs
Muscles are made up of bundles of long cells that either contract (shorten) or relax (lengthen) when triggered by the nervous system. The most common type of muscles are those connected to the bones of the skeleton. They pull on the bones when they contract, causing actions like the movement of this arm. Because muscles cannot push, they have to work in pairs – one muscle to pull the arm upwards and another to pull it back down.

Rotating the arm
As well as pulling the lower arm towards it, the biceps can also rotate the forearm so that the palm of the hand faces upwards.

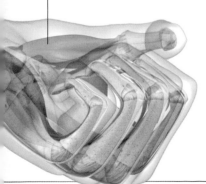

Plant movement
Like animals, plants move to make the most of their environment. The shoot tips of plants are especially sensitive to light and can slowly bend towards a light source. A chemical called auxin (which regulates growth) encourages the shadier side of the shoot to grow more, bending the plant towards the Sun.

Auxin (pink) produced in the shoot tip spreads down through the shoot, making it grow upwards.

When light shines from one direction, auxin moves to the shadier side.

On the shadier side, the auxin stimulates the plant cells to grow bigger, so that the shoot bends towards the light.

Support structures
Animals have a skeleton to support their bodies and protect their soft organs. This is especially important for large land-living animals that are not supported by water. Skeletons also provide a firm support for contracting muscles, helping animals to have the strength to move around.

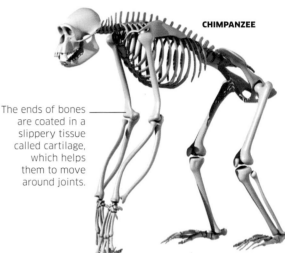

CHIMPANZEE

The ends of bones are coated in a slippery tissue called cartilage, which helps them to move around joints.

Endoskeleton
Vertebrate animals – including fish, amphibians, reptiles, birds, and mammals – have a hard internal skeleton within their bodies. The muscles surround the skeleton and pull on its bones.

The skeleton is thinner and more flexible around the joints.

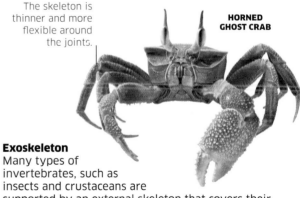

HORNED GHOST CRAB

Exoskeleton
Many types of invertebrates, such as insects and crustaceans are supported by an external skeleton that covers their body like a suit of armour, with muscles inside. Exoskeletons cannot grow with the rest of the body, so must be periodically shed and replaced.

MOON JELLYFISH

Muscles in a jellyfish contract around a layer of thin jelly that keeps its body firm.

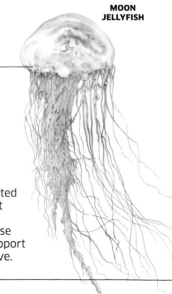

Hydroskeleton
Some kinds of soft-bodied animals, such as sea anemones and earthworms, are supported by internal pouches that stay firm because they are filled with fluid. These water-filled pouches support the muscles as they move.

Pulsating

A jellyfish moves neither forwards or backwards, but instead rises and falls in the water. It has a ring of muscles around the rim of its bell. When these contract, the bell closes slightly like a drawstring bag, forcing water out and shooting the animal upwards.

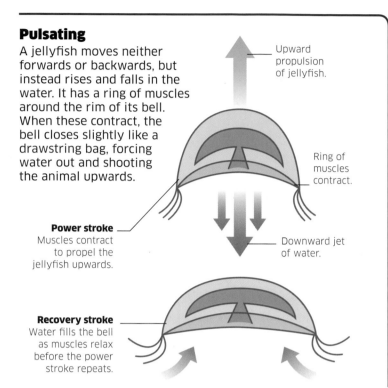

Upward propulsion of jellyfish.

Ring of muscles contract.

Power stroke
Muscles contract to propel the jellyfish upwards.

Downward jet of water.

Recovery stroke
Water fills the bell as muscles relax before the power stroke repeats.

Getting around

Whether over land, underwater, or in the air, animals can move themselves around in extraordinary ways when all their muscles work together.

Although all living things move parts of their body to an extent, only animals can truly "locomote". This is when the entire body moves to a different location. Some animals do it without any muscle power at all – riding on ocean currents, or getting blown by the wind. But most animals locomote under their own steam. They do so for many different reasons: to find food or a mate, or to escape from predators. Some animals migrate over enormous distances from season to season, or even from day to day.

Tiny, deep ocean pygmy sharks grow no bigger than 20 cm (8 in), but each night **swim 1.5 km (1 mile) up to the surface and back** in order to feed.

Tiger beetles have long legs to give them speed.

Tiger beetle

Predatory tiger beetles are fast sprinters. Like all insects, they have six legs, and when running they lift three simultaneously, leaving three in contact with the ground. However, their big eyes cannot keep up with their speed, meaning their vision is blurred every time they run.

Running

An animal that moves over land needs its muscles to pull against a strong supporting framework. It also needs good balance to stay upright, meaning that its muscles and skeleton must work together with the nervous system. Some animals move slowly, even when in a hurry, but others are born to run. The fastest runners not only have powerful muscles to move their limbs more quickly, but also take much longer strides.

A cheetah can accelerate to **100 km/h (62 mph)** in just three seconds.

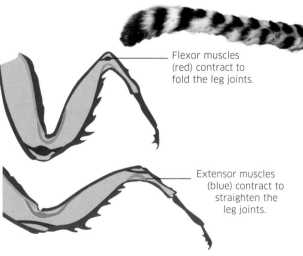

Flexor muscles (red) contract to fold the leg joints.

Extensor muscles (blue) contract to straighten the leg joints.

Multi-jointed leg

Arthropods, including insects, spiders, and crabs, have multi-jointed legs that carry an armour-like outer skeleton, with their muscles attached on the inside. Their muscles work in pairs around each joint: one to flex (bend) and the other to extend.

A flexible spine helps the cheetah's body bend up and down when sprinting.

Long legs deliver long strides.

All four feet are airborne at least twice for each stride.

Cheetah

The cheetah accelerates faster than any other land animal, but it is not just its long legs that help it pick up speed. Humans run on the flats of their feet, but in cats the toes bear the weight – effectively lengthening the limb. The cheetah's flexible backbone helps make its legs swing wider, adding 10 per cent to its stride.

The strongest jumper is the froghopper bug, which leaps a distance 70 times its body size.

120 km/h (75 mph) – the speed of a peregrine falcon as it dives for prey in the sky. It is the fastest animal of all.

173

Burrowing

Life underground comes with special challenges. Burrowers need the strength to dig through soil to create a passage, and the ability to crawl through small openings. Moles use their feet like shovels to claw back the soil, but earthworms bulldoze their way through with their bodies.

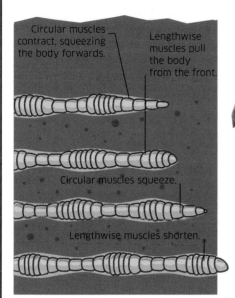

Circular muscles contract, squeezing the body forwards.

Lengthwise muscles pull the body from the front.

Circular muscles squeeze.

Lengthwise muscles shorten.

Earthworm
An earthworm has two sets of muscles. One set encircles the body and squeezes to push it forwards, like toothpaste from a tube. The other pulls the body forwards.

Peach-faced lovebird
These birds are supremely adapted for powered flight. They have massive flight muscles to flap their wings, hollow bones to make them lightweight, and feathers to help with streamlining.

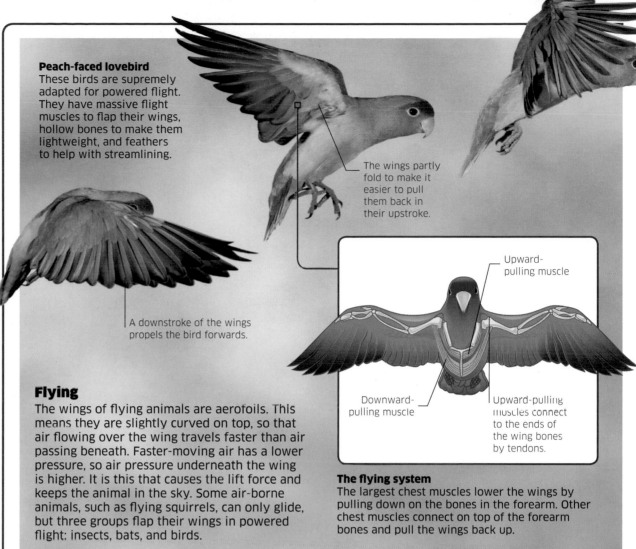

The wings partly fold to make it easier to pull them back in their upstroke.

A downstroke of the wings propels the bird forwards.

Flying
The wings of flying animals are aerofoils. This means they are slightly curved on top, so that air flowing over the wing travels faster than air passing beneath. Faster-moving air has a lower pressure, so air pressure underneath the wing is higher. It is this that causes the lift force and keeps the animal in the sky. Some air-borne animals, such as flying squirrels, can only glide, but three groups flap their wings in powered flight: insects, bats, and birds.

Upward-pulling muscle

Downward-pulling muscle

Upward-pulling muscles connect to the ends of the wing bones by tendons.

The flying system
The largest chest muscles lower the wings by pulling down on the bones in the forearm. Other chest muscles connect on top of the forearm bones and pull the wings back up.

Swimming

Water is thicker than air, so it exerts a bigger force called drag against any animal that moves through it. Swimming animals reduce drag by being streamlined. Even though marine animals, such as fish and dolphins are only distantly related, they both have similar body shapes, to better propel themselves through the water.

Swimming fish
Fish have blocks of muscle in the sides of their body. These contract to bend the body in an "S" shape, sweeping the tail from side to side and propelling the fish forwards.

Sailfish
The enormous fin of a sailfish helps to steady its body – letting it get close to prey undetected. However, when the sail is lowered, it gives chase faster than any other fish in the ocean.

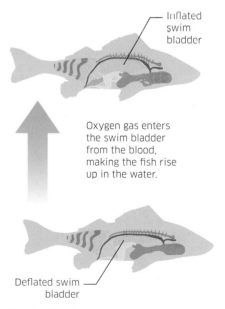

Inflated swim bladder

Oxygen gas enters the swim bladder from the blood, making the fish rise up in the water.

Deflated swim bladder

Controlling buoyancy
Fish are heavier than water, but most bony fish have a gas-filled chamber –the swim bladder – for staying buoyant when swimming. By controlling the volume of gas inside the swim bladder, fish can rise or sink through different water levels.

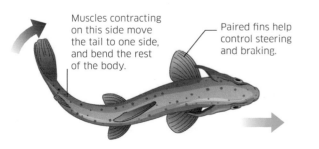

Muscles contracting on this side move the tail to one side, and bend the rest of the body.

Paired fins help control steering and braking.

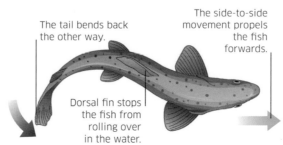

The tail bends back the other way.

The side-to-side movement propels the fish forwards.

Dorsal fin stops the fish from rolling over in the water.

Plant reproduction

Despite being rooted in the ground, plants work hard to ensure the survival of their species. With the help of wind, water, and animals, they fertilize one another and disperse their seeds far and wide.

Flowers are the reproductive organs of most kinds of plants and contain both male and female cells. The male cells – encased in dusty pollen grains – fertilize eggs in the flower's female parts. Each tiny young plant produced is then enclosed inside a seed: a survival capsule that protects its contents until they are ready to germinate.

Carpenter bees visit the flowers to collect sugary nectar – an energy-rich food.

Stamen
Yellow stamens, which produce dust-like pollen grains, are the male parts of the plant.

1 Flowering
The flower's vibrant purple stripes guide a carpenter bee to the nectar glands at its centre. Other plants with less attractive flowers may instead scatter their pollen on the wind.

2 Pollination
Yellow stamens brush the insect's hairy body with pollen, which the bees carry with them to the purple club-like stigmas of another plant in the species.

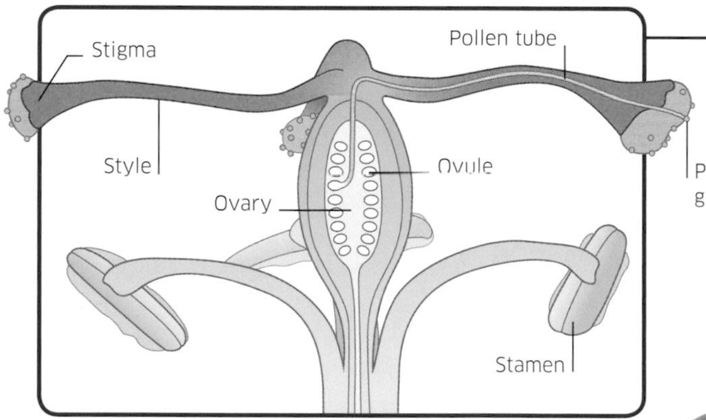

Stigma

Pollen tube

Style

Ovule

Ovary

Pollen grain

Stamen

3 Fertilization
After landing on the stigma, the pollen grains sprout microscopic tubes to carry their male cells down the style to reach the female eggs. Each fertilized egg then grows into an embryo, nestled inside a white capsule called an ovule.

After fertilization, the petals of a flower shrivel and fall off.

Reproduction partnerships

Like many kinds of plants, the passion vine from South America relies on animals to help it reproduce. Large hairy carpenter bees in search of sweet nectar carry pollen from flower to flower, while birds with a taste for fruit – here the great kiskadee – spread the seeds.

4 Fruiting
When fertilized, the base of the flower begins to develop into a fleshy fruit. The ovules embedded inside harden to form seeds.

Many kinds of flowers are **pollinated by insects**, but others use **birds or even bats**.

The **seeds of fir trees** and related plants develop from **cones instead of flowers**.

175

6 Germination
If seeds land on moist ground, the embryos inside them start to grow and the seeds germinate. Roots grow down to absorb water and minerals, while shoots sprout upwards to make leaves.

A new shoot emerges from the split seed capsule.

Leaves spring up as the plant develops.

Stigma
The purple stigmas are female parts of the flower, which collect the pollen grains.

Style
A style connects each stigma to the ovary at its base.

5 Seed dispersal
The fruit turns orange and gets sweeter as it ripens. This attracts fruit-eaters, such as the great kiskadee, which consume the fruit and scatter the plant's seeds in their droppings.

Birds can spread seeds far away from the original plant, but they are not the only way these tiny capsules travel. Other seed species may be carried by wind, or water.

Asexual reproduction

Many plants can reproduce asexually – meaning without producing male and female sex cells. Some develop side shoots, or runners, that split away into new plants. A few grow baby plants on their leaves.

Tiny new plants growing on the leaf of a hen-and-chicken fern fall off to produce entirely new ferns.

Reproducing by spores

Mosses and ferns do not produce flowers and seeds, but scatter spores instead. Spores are different from seeds, as they contain just a single cell rather than a fertilized embryo. These cells grow into plants with reproductive organs, which must fertilize each other to develop into mature plants that can produce a new generation of spores.

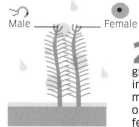

Spore capsule

1 Scattering spores
Fully-grown moss shoots release countless single-celled spores from spore capsules. These are carried by the wind, landing where each can grow into a new plant.

Male — Female

2 Sex organs develop
Landing on moist ground, the spores grow into tiny, leafy shoots with microscopic sex organs. Male organs produce sperm, and female organs produce eggs.

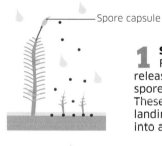

3 Fertilization
Falling raindrops allow swimming sperm cells to reach the eggs held inside the female sex organs, where they fertilize them.

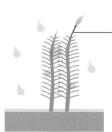

Spore-producing shoot

4 Spore capsule grows
Each fertilized egg grows into a new spore-producing shoot with a spore capsule, ready to make more spores and repeat the life cycle.

Producing young

The drive to reproduce is one of the most basic instincts in all animals. Many species devote their entire lives to finding a mate and making new young.

The most common way for animals to reproduce is through sexual reproduction – where sperm cells produced by a male fertilize egg cells produced by a female. The fertilized egg then becomes an embryo that will slowly grow and develop into a new animal. Many underwater animals release their sperm and eggs together into open water, but land animals must mate so that sperm are passed into the female's body and can swim inside it to reach her eggs.

Laying eggs on land

In some land animals, such as birds and reptiles, eggs are fertilized inside the mother's body and then laid – usually into a nest. These eggs have a hard, protective shell that encases the embryo inside and stops it from drying out. They also contain a big store of food – the yolk – which nourishes the embryo as it develops. It can take weeks or even months before the baby is big enough to hatch and survive in the world outside.

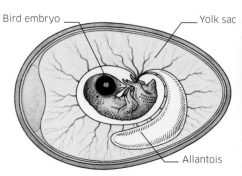

Bird embryo   Yolk sac

Allantois

Inside a bird's egg
The shell of a bird's egg lets in air to help the embryo breathe. The yolk sac provides nutrients as it grows into a chick, while another sac (the allantois) helps collect oxygen and waste.

Giving birth to live young

Except for a few egg-laying species (called monotremes), all mammals give birth to live young. The mother must support the growing embryos inside her body – a demanding task that may involve her taking in extra nutrients. The babies grow in a part of the mother's body called the uterus, or womb, where a special organ called a placenta passes them food and oxygen.

A new generation of mice
Some mammals, such as humans, usually give birth to one baby at a time. Others have large litters – like mice, which can produce up to 14 babies at one time. Each one starts as a fertilized egg, grows into an embryo, and then is born just three weeks later.

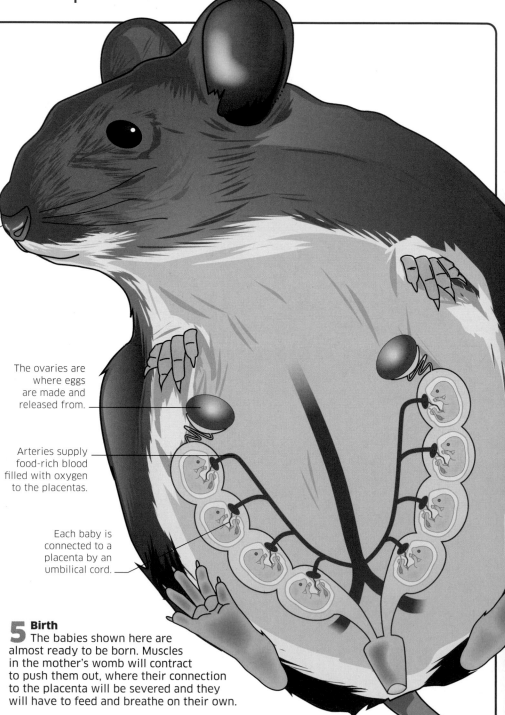

The ovaries are where eggs are made and released from.

Arteries supply food-rich blood filled with oxygen to the placentas.

Each baby is connected to a placenta by an umbilical cord.

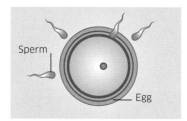

Sperm

Egg

1 Fertilization
When a male mouse mates with a female, thousands of sperm enter her body and swim to her eggs. The first to arrive penetrates an egg and fertilizes it.

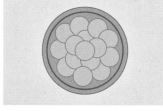

2 Embryo forms
The fertilized egg cell now contains a mixture of genes from the sperm and the egg. It divides multiple times to form a microscopic ball of cells called an embryo.

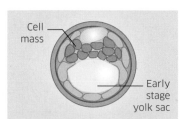

Cell mass

Early stage yolk sac

3 Implantation
The embryo becomes a hollow ball. A cell mass on one side will become the mouse's body. The ball travels deep into the womb to embed in its wall, an event called implantation.

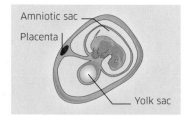

Amniotic sac

Placenta

Yolk sac

4 Placenta grows
The baby mouse begins to form and gets nutrients from first a temporary yolk sac, and then a placenta. A fluid-filled bag called the amniotic sac cushions the embryo.

5 Birth
The babies shown here are almost ready to be born. Muscles in the mother's womb will contract to push them out, where their connection to the placenta will be severed and they will have to feed and breathe on their own.

300 million eggs can be produced by the ocean sunfish at one time – more than any other back-boned animal.

Female seahorses lay their eggs inside a pouch on the male's body, so that it is the father that gives birth to them.

177

Laying eggs in water

Fish fertilize their eggs externally, so the females lay unfertilized eggs directly into the water. Instead of having hard shells, fish eggs are usually coated in a soft jelly that will cushion and protect the developing embryos. Most fish do not wait around to see the embryos develop, but simply scatter lots of floating eggs and swimming sperm and leave the outcome to chance. However, some species, such as clownfish, carefully tend to their developing babies.

1 Laying and fertilization
A female clownfish lays her eggs onto a hard surface. The male then releases his sperm to fertilize them.

2 Caring for the eggs
During the week it takes for them to hatch, the father guards over the eggs, using his mouth to clean them.

3 Hatching
Tiny babies, called fry, break out of the eggs. They grow quickly, feeding on nutrients in their yolk sac.

Investing in babies

All animals spend a lot of time and effort in breeding, but they invest this energy in different ways. Some – such as many insects and most fish – produce thousands of eggs at once, and a few even die after breeding. Others produce just one baby at a time, but spend a lot of time caring for each one.

Breeding lifetimes

Animals must have fully grown reproductive systems before they can breed, and some can take years to develop these. While some animals breed often throughout their long lives, shorter-lived species make up for their limited lifespans by producing many babies each time.

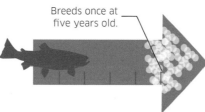

Breeds once at five years old.

Sockeye salmon
Salmon swim from the ocean into rivers to scatter millions of eggs. They die after this huge effort, so can only breed once.

Starts breeding at three years old.

Can breed until 10 years old.

Common toad
Toads can keep breeding for seven years of their adult life, and each year produce thousands of eggs.

The pregnancy of an elephant lasts for 22 months, the longest time of any mammal.

African elephant
It takes so long to rear an elephant calf that elephants only manage it every few years. However, they continue to reproduce for many decades.

Starts breeding at 20 years old.

Continues breeding until 60 years old.

Parental care

The best way to ensure that babies survive is to give them good care when they are at their most vulnerable, but animal parents vary a lot in their degree of devotion. Many invertebrates give limited parental care or none at all. But mammal babies may be nurtured by their parents for many years.

Newborn kangaroos live in a pouch in their mother's bodies, where they continue to grow and develop.

Coral
Adult coral provide no parental care. Young microscopic stages of coral – called larvae – must fend for themselves in the open ocean, where most will get eaten by predators.

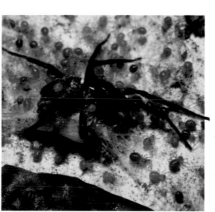

Black lace weaver spider
This spider mother makes the ultimate sacrifice for her babies. After laying more eggs for her young to eat, she encourages them to bite her. This stirs their predatory instincts, and they eat her.

Orang-utan
Childhood for this tree-living ape lasts well into the teenage years, just like in humans. During this time, the young will stick close to their mother for protection and learn vital survival skills from her.

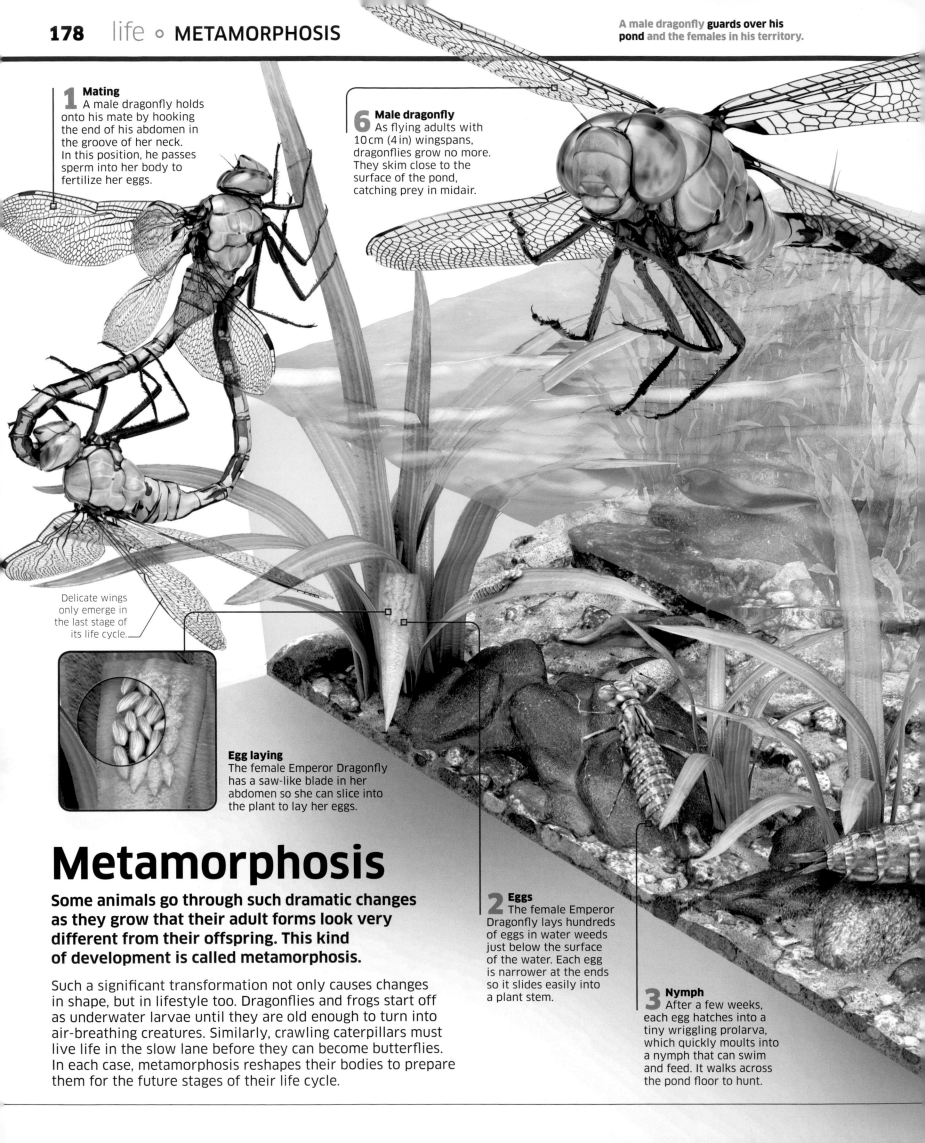

A male dragonfly **guards over his pond** and the females in his territory.

1 Mating
A male dragonfly holds onto his mate by hooking the end of his abdomen in the groove of her neck. In this position, he passes sperm into her body to fertilize her eggs.

6 Male dragonfly
As flying adults with 10 cm (4 in) wingspans, dragonflies grow no more. They skim close to the surface of the pond, catching prey in midair.

Delicate wings only emerge in the last stage of its life cycle.

Egg laying
The female Emperor Dragonfly has a saw-like blade in her abdomen so she can slice into the plant to lay her eggs.

Metamorphosis

Some animals go through such dramatic changes as they grow that their adult forms look very different from their offspring. This kind of development is called metamorphosis.

Such a significant transformation not only causes changes in shape, but in lifestyle too. Dragonflies and frogs start off as underwater larvae until they are old enough to turn into air-breathing creatures. Similarly, crawling caterpillars must live life in the slow lane before they can become butterflies. In each case, metamorphosis reshapes their bodies to prepare them for the future stages of their life cycle.

2 Eggs
The female Emperor Dragonfly lays hundreds of eggs in water weeds just below the surface of the water. Each egg is narrower at the ends so it slides easily into a plant stem.

3 Nymph
After a few weeks, each egg hatches into a tiny wriggling prolarva, which quickly moults into a nymph that can swim and feed. It walks across the pond floor to hunt.

5 Dragonfly emerges
Just before its final moult, a nymph climbs up a plant out of the water. This time a dragonfly emerges from its skin.

It can take three hours for the insect's wings to harden so it can fly.

4 Large nymph
The nymph passes through several more moults, growing each time. All insects must regularly moult their outer skin because this strong casing cannot expand as they grow.

Hunting tools
An Emperor Dragonfly nymph has a massive clawed lower "lip", which it can shoot out in just a fraction of a second to grab its prey. The nymphs can grow over 5 cm (2 in) long – large enough to grab large prey such as fish.

Complete metamorphosis

Along with many other insects, a butterfly undergoes a different kind of metamorphosis to a dragonfly. Its larva is a caterpillar, a leaf-eating creature that has no resemblance to the adult form at all. It changes into a flying butterfly in a single transformation event. This process is different to incomplete metamorphosis, where the multiple larval forms are smaller versions of the adult.

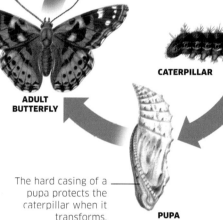

EGG

CATERPILLAR

The hard casing of a pupa protects the caterpillar when it transforms.

PUPA

ADULT BUTTERFLY

Amphibian life cycle

Amphibians grow more gradually than insects because they do not need to moult. Tiny wriggling tadpoles – with gills for breathing underwater – hatch from frogspawn and then take weeks or months to get bigger and turn into air-breathing frogs. During this time, they steadily grow their legs and their tails get absorbed back into their bodies.

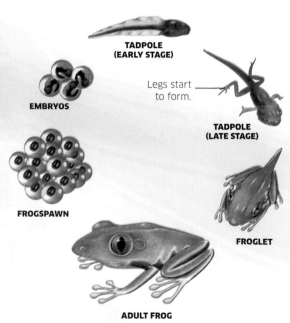

TADPOLE (EARLY STAGE)

EMBRYOS

Legs start to form.

TADPOLE (LATE STAGE)

FROGSPAWN

FROGLET

ADULT FROG

Packaging the information

Inside the nucleus of every cell in the human body are 46 molecules of DNA, carrying all the information needed to build and maintain a human being. Each molecule is shaped like a twisted ladder – named a double helix – and packaged up into a bundle called a chromosome. Genetic information is carried by the sequence of different chemical units, known as bases, that make up the "rungs" of the ladder.

Genetics and DNA

The characteristics of a living thing – who we are and what we look like – are determined by a set of instructions carried inside each of the body's cells.

Instructions for building the body and keeping it working properly are held in a substance called DNA (deoxyribonucleic acid). The arrangement of chemical building blocks in DNA determines whether a living thing grows into an oak tree, a human being, or any other kind of organism. DNA is also copied whenever cells divide, so that all the cells of the body carry a set of these vital genetic instructions. Half of each organism's DNA is also passed on to the next generation in either male or female sex cells.

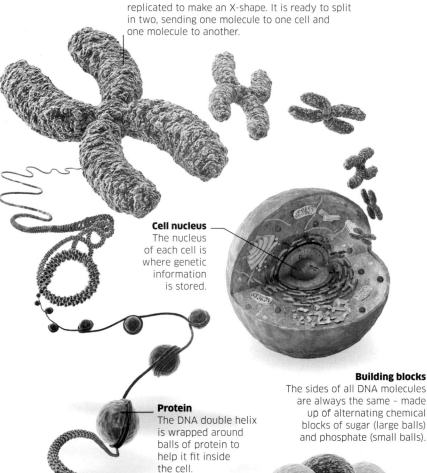

Chromosome
A tightly packed mixture of protein and DNA forms a chromosome. Each chromosome contains one long molecule of DNA. The DNA in this chromosome has replicated to make an X-shape. It is ready to split in two, sending one molecule to one cell and one molecule to another.

Cell nucleus
The nucleus of each cell is where genetic information is stored.

Protein
The DNA double helix is wrapped around balls of protein to help it fit inside the cell.

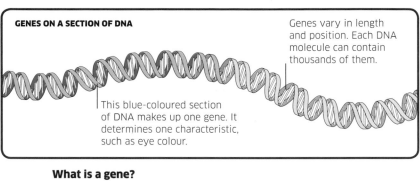

GENES ON A SECTION OF DNA

Genes vary in length and position. Each DNA molecule can contain thousands of them.

This blue-coloured section of DNA makes up one gene. It determines one characteristic, such as eye colour.

What is a gene?

The information along DNA is arranged in sections called genes. Each gene has a unique sequence of bases. This sequence acts as a code to tell the cell to make a specific protein, which, in turn, affects a characteristic of the body.

Building blocks
The sides of all DNA molecules are always the same – made up of alternating chemical blocks of sugar (large balls) and phosphate (small balls).

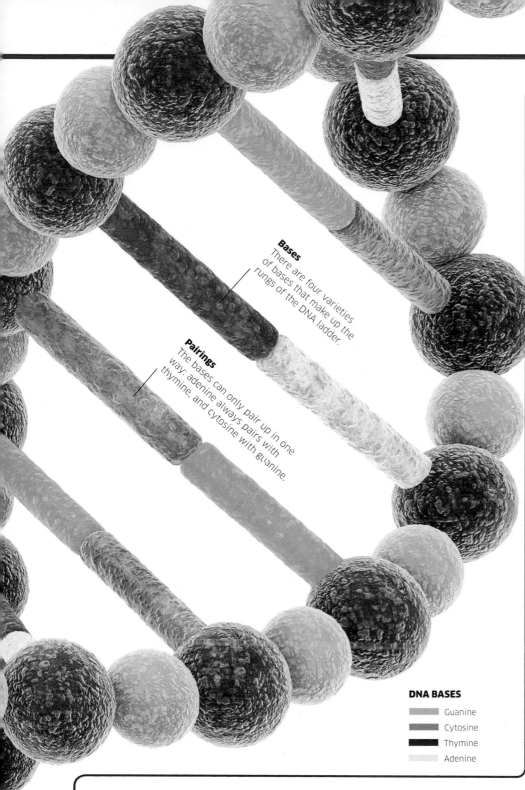

Bases
There are four varieties of bases that make up the rungs of the DNA ladder.

Pairings
The bases can only pair up in one way: adenine always pairs with thymine, and cytosine with guanine.

DNA BASES

Guanine
Cytosine
Thymine
Adenine

DNA replication

Cells replicate themselves by splitting in two. Therefore, all the instructions held in DNA must be copied before a cell divides, so each new cell will have a full set. The DNA does this by splitting into two strands. Each of these then provides a template for building a new double helix.

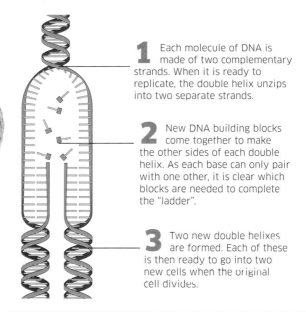

1 Each molecule of DNA is made of two complementary strands. When it is ready to replicate, the double helix unzips into two separate strands.

2 New DNA building blocks come together to make the other sides of each double helix. As each base can only pair with one other, it is clear which blocks are needed to complete the "ladder".

3 Two new double helixes are formed. Each of these is then ready to go into two new cells when the original cell divides.

What gets inherited?

Many human features, such as eye colour, hair colour, and blood group, are due to particular genes. Different varieties of genes, called alleles, determine variation in these characteristics. Other characteristics, such as height, are affected by many genes working together, but also by other factors, such as diet.

Genes
Some characteristics are only inherited from parents.

EYE COLOUR

EARLOBE SHAPE

Genes and environment
Other characteristics are influenced by genes and the environment.

AGE WHEN HAIR TURNS GREY

EYESIGHT

How inheritance works

Most organisms have two lots of each kind of gene – one from each parent. Many genes have two or more variations, called alleles, so the genes an animal inherits from its mother and father may be identical or different. When two animals, such as rabbits, reproduce, there are many different combinations of alleles their offspring can receive. Some alleles are dominant (like those for brown fur), and when a baby rabbit has two different alleles it will have the characteristic of the dominant allele. Other alleles are recessive, and babies will only have the characteristic they determine if they have two of them. This explains why some children inherit physical characteristics not seen in their parents.

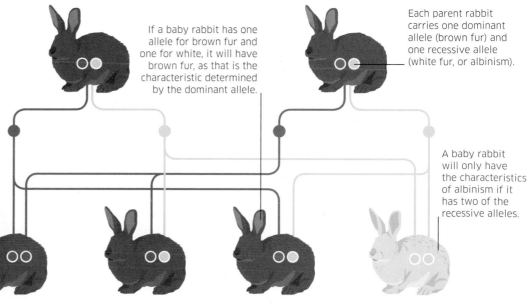

If a baby rabbit has one allele for brown fur and one for white, it will have brown fur, as that is the characteristic determined by the dominant allele.

Each parent rabbit carries one dominant allele (brown fur) and one recessive allele (white fur, or albinism).

A baby rabbit will only have the characteristics of albinism if it has two of the recessive alleles.

Mosquitoes change habitats throughout their lives, dwelling in ponds as larvae, but taking to the air when fully grown.

A place to live

Every form of life – each species of plant, animal, or microbe – has a specific set of needs that means it can only thrive in suitable places.

Habitats are places where organisms live. A habitat can be as small as a rotting log or as big as the open ocean, but each one offers a different mixture of conditions that suits a particular community of species. Here, the inhabitants that are adapted to these conditions – to the habitat's climate, food, and all other factors – can grow and survive long enough to produce the next generation.

Life between the tides

Nowhere can the diversity between habitats be seen better than where the land meets the sea. Conditions vary wildly on a rocky shore – from the submerged pools of the lower levels to the exposed land higher up. As the tide moves in and out daily, many species must be adapted to a life spent partly in the open air and partly underwater.

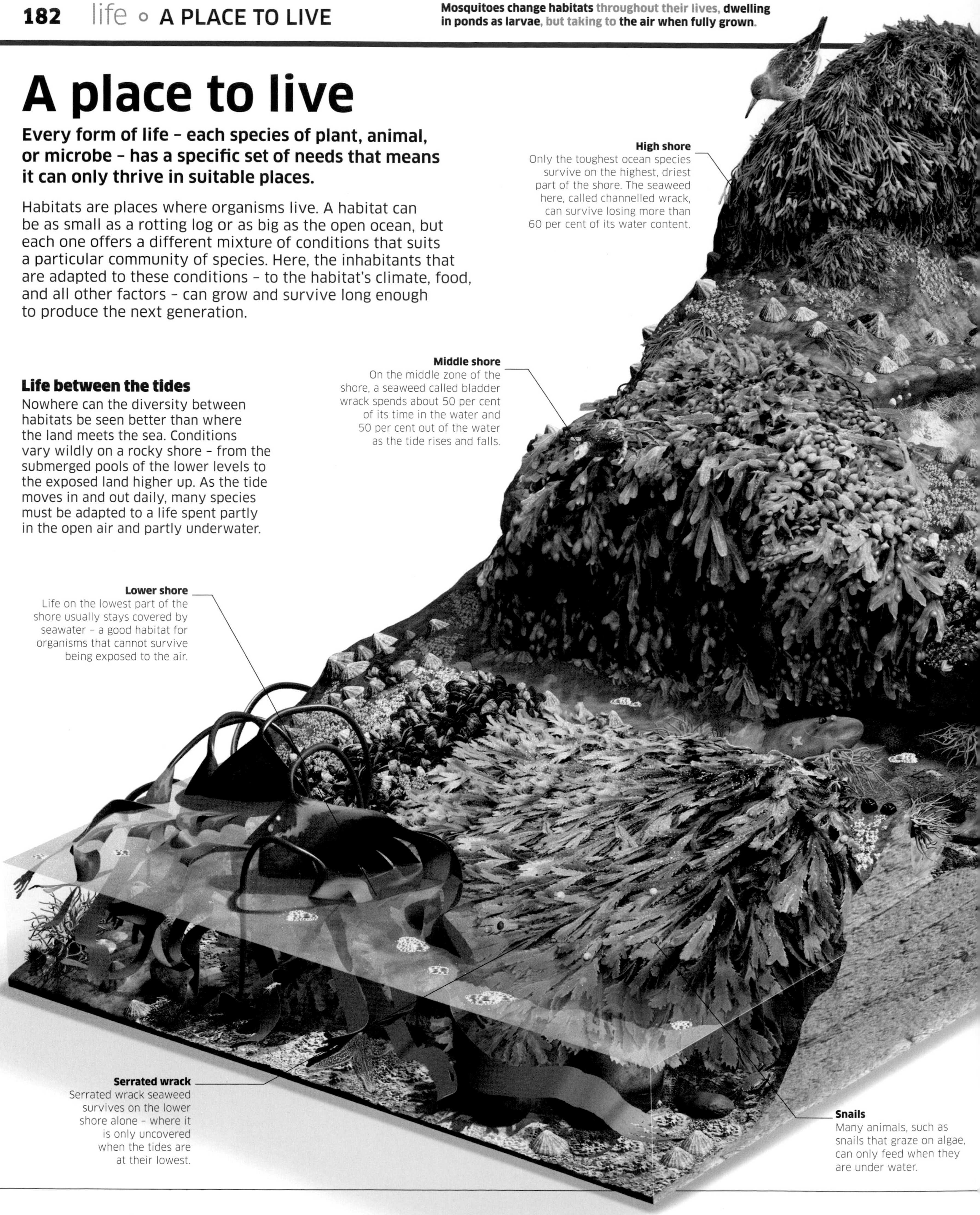

High shore
Only the toughest ocean species survive on the highest, driest part of the shore. The seaweed here, called channelled wrack, can survive losing more than 60 per cent of its water content.
Middle shore
On the middle zone of the shore, a seaweed called bladder wrack spends about 50 per cent of its time in the water and 50 per cent out of the water as the tide rises and falls.
Lower shore
Life on the lowest part of the shore usually stays covered by seawater – a good habitat for organisms that cannot survive being exposed to the air.
Serrated wrack
Serrated wrack seaweed survives on the lower shore alone – where it is only uncovered when the tides are at their lowest.
Snails
Many animals, such as snails that graze on algae, can only feed when they are under water.

Many organisms have **urban habitats** – from secretive **house mice** to **large leopards** that roam free in the city of Mumbai, India.

Microscopic **bacteria** are found in every community. There could be **thousands of species** of bacteria in just **a single teaspoon of soil.**

183

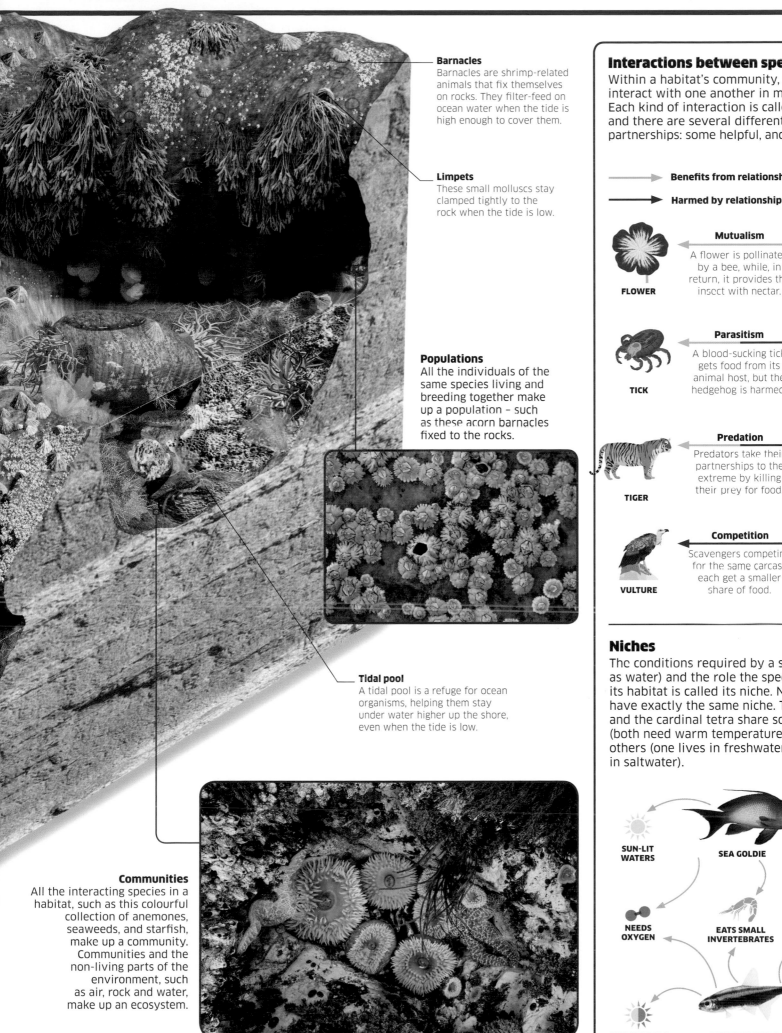

Barnacles
Barnacles are shrimp-related animals that fix themselves on rocks. They filter-feed on ocean water when the tide is high enough to cover them.

Limpets
These small molluscs stay clamped tightly to the rock when the tide is low.

Populations
All the individuals of the same species living and breeding together make up a population – such as these acorn barnacles fixed to the rocks.

Tidal pool
A tidal pool is a refuge for ocean organisms, helping them stay under water higher up the shore, even when the tide is low.

Communities
All the interacting species in a habitat, such as this colourful collection of anemones, seaweeds, and starfish, make up a community. Communities and the non-living parts of the environment, such as air, rock and water, make up an ecosystem.

Interactions between species

Within a habitat's community, species interact with one another in many ways. Each kind of interaction is called a symbiosis, and there are several different kinds of partnerships: some helpful, and some harmful.

→ Benefits from relationship

→ Harmed by relationship

Mutualism
A flower is pollinated by a bee, while, in return, it provides the insect with nectar.

FLOWER   BEE

Parasitism
A blood-sucking tick gets food from its animal host, but the hedgehog is harmed.

TICK   HEDGEHOG

Predation
Predators take their partnerships to the extreme by killing their prey for food.

TIGER   GOAT

Competition
Scavengers competing for the same carcass each get a smaller share of food.

VULTURE   HYENA

Niches

The conditions required by a species (such as water) and the role the species plays in its habitat is called its niche. No two species have exactly the same niche. The sea goldie and the cardinal tetra share some conditions (both need warm temperatures), but not others (one lives in freshwater, the other in saltwater).

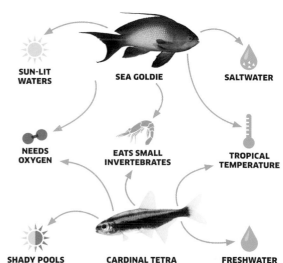

SUN-LIT WATERS   SEA GOLDIE   SALTWATER

NEEDS OXYGEN   EATS SMALL INVERTEBRATES   TROPICAL TEMPERATURE

SHADY POOLS   CARDINAL TETRA   FRESHWATER

184 life ∘ **HABITATS AND BIOMES**

Coral reefs cover less than **1 per cent of the ocean floor,** but are **home to more than a quarter of all known species.**

Oceanic zones

Covering nearly three-quarters of the Earth's surface, and reaching down to 11 km (9 miles) at their deepest point, the oceans make up the biggest biome by volume. All life here lives submerged in salty marine waters, but conditions vary enormously from the coastlines down to the ocean's bottom.

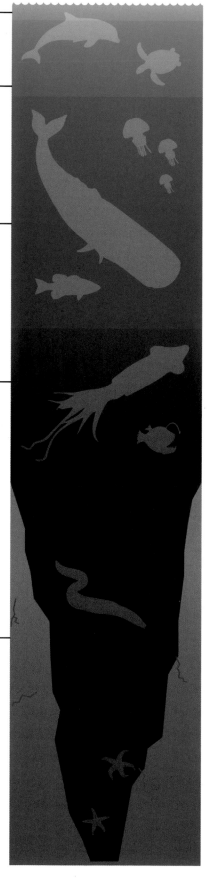

Sunlit zone
(0–200 m/0–650 ft)
Bright sunlight provides energy for ocean food chains that start with algae.

Twilight zone
(200–1,000 m/650–3,280 ft)
Sunlight cannot penetrate far into the ocean, so as depth increases, conditions are too dark for algae, but animals thrive.

Midnight zone
(1,000–4,000 m/ 3,280–13,000 ft)
Animals find different ways of surviving in the dark ocean depths. Many use bioluminescence: they have light-producing organs to help them hunt for food or avoid danger.

Abyssal zone
(4,000–6,000 m/ 13,000–19,650 ft)
Near the ocean floor, water pressure is strong enough to crush a car and temperatures are near freezing. Most food chains here are supported by particles of dead matter raining down from above.

Hadal zone
(6,000–11,000 m/ 19,650–36,000 ft)
The ocean floor plunges down into trenches that form the deepest parts of the ocean. But even here there is life – with a few kinds of fishes diving down to 8,000 m (26,000 ft) and invertebrates voyaging deeper still.

Biomes

Places exposed to similar sets of conditions – such as temperature or rainfall – have similar-looking habitats, even when they are as far apart as America and Asia. These habitat groups are called biomes. Over continents and islands, they include tundra, deserts, grasslands, forests – and freshwater lakes and rivers.

Tundra
Where land is close to the poles, conditions are so cold that the ground is permafrost – meaning it is frozen throughout the year. Here, trees are sparse or cannot grow at all, and the thin vegetation is made up of grasses, lichens, and small shrubs.

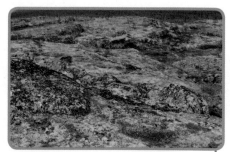

Taiga
The largest land biome is a broad belt of coniferous forest that encircles the world below the Arctic tundra. Conifers, pines, and related trees have needle-like leaves that help them survive low temperatures. They are evergreen – so retain their tough foliage even in the coldest winters.

Temperate forest
The Earth's temperate zones are between the cold polar regions and the tropics around the equator. Many of the forests that grow in these seasonal regions are deciduous: they produce their leaves during the warm summers, but lose them in the cold winters.

Temperate grassland
Where the climate is too dry to support forests but too wet for desert, the land is covered with grassland, a habitat that supports a wide range of grazing animals. Temperate grasslands experience seasonal changes in temperature, but stay green throughout the year.

Tropical dry and coniferous forest
Some tropical regions have pronounced dry seasons that can last for months. Here, many kinds of trees drop their leaves in times of drought. Others have adaptations that help them to stay evergreen. In places, the forests are dominated by conifers with drought-resistant leaves.

Extremophiles are organisms that live in extreme habitats – such as bacteria that thrive at 120°C (248°F) around volcanic vents.

Habitats change: Antarctica is covered in snow and ice today, but 52 million years ago rainforests grew there.

185

Freshwater
Rainfall collecting in rivers and lakes creates freshwater habitats. Aquatic plants grow in their shallows and animals swim in the open water or crawl along their muddy or stony bottoms. Where rivers meet the sea, water is affected by the oceans' saltiness.

Habitats and biomes

Around the Earth, plants, animals, and other organisms live in habitats that are as different as the driest, most windswept deserts and the deepest, darkest oceans.

Conditions vary from one part of the world to another, and they have a big effect on the kinds of living things that can survive together in any place. The freezing cold poles experience a winter of unbroken darkness for half the year, while the equator basks in tropical temperatures year-round. And the world of the oceans reaches from the sunlit surfaces down into the dark abyss.

Mediterranean woodland
A Mediterranean-type climate has hot, dry summers and wet, mild winters. It is commonest where lands in the temperate zone are influenced by mild ocean air. Its forests are dominated by trees, such as eucalyptus, that are sclerophyll, meaning they have leathery, heat-resistant leaves.

Montane grassland and shrubland
Temperatures drop with increasing altitude, so the habitat changes in mountain regions. Forests give way to grassland on exposed slopes, which are then replaced with sparser vegetation, called montane tundra, higher up.

Tropical rainforest
Where temperature, rainfall, and humidity remain high all year round, the Earth is covered with tropical rainforest. These are the best conditions for many plants and animals to grow, and they have evolved into more different species than in any other land biome.

Desert
In some parts of the world – in temperate or tropical regions – the land receives so little rainfall that conditions are too dry for most grasses and trees. In arid places with hot days and cold nights, succulent plants survive by storing water in roots, stems, or leaves.

Tropical grassland
Grasslands in the tropics support some of the largest, most diverse gatherings of big grazing animals anywhere on Earth. Unlike most plants, grasses grow from the base of their leaves, so thrive even when vast numbers of grazers eat the top of their foliage.

Cycles of matter

Many of Earth's crucial materials for life are constantly recycled through the environment.

All the atoms that make up the world around us are recycled in one way or another. Chemical reactions in living things, such as photosynthesis and respiration, drive much of this recycling. These processes help pass important elements like carbon and nitrogen between living things, the soil, and the atmosphere.

Nitrogen gas makes up about **two-thirds of Earth's atmosphere.**

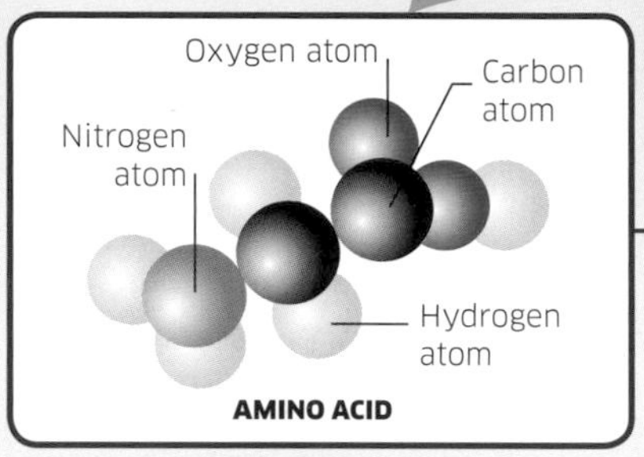

Oxygen atom

Carbon atom

Nitrogen atom

Hydrogen atom

AMINO ACID

Nitrogen in molecules
Molecules containing nitrogen are used by plants, animals, and bacteria, such as this amino acid. It helps with growth and other vital functions.

Plants use nitrate from the roots to make food. When a leaf falls, it still contains this nitrogen.

NITROGEN MOLECULES

The nitrogen cycle
Nitrogen exists in many forms inside living things, including in DNA, proteins, and amino acids. Animals and many bacteria obtain their nitrogen by feeding on other organisms – dead or alive. Plants absorb it as a mineral called nitrate – a chemical that gets released into the soil through the action of the bacteria.

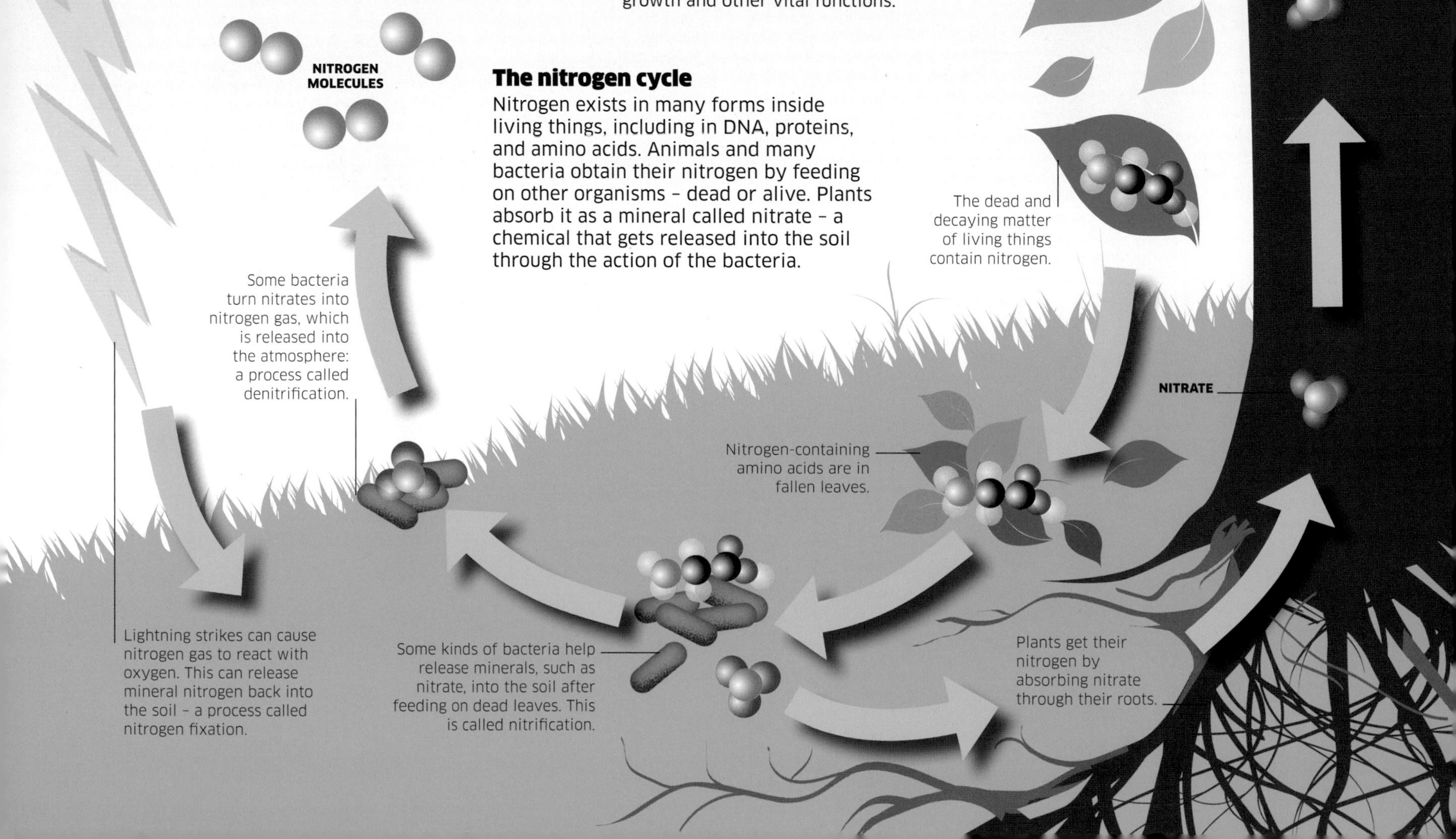

The dead and decaying matter of living things contain nitrogen.

Some bacteria turn nitrates into nitrogen gas, which is released into the atmosphere: a process called denitrification.

NITRATE

Nitrogen-containing amino acids are in fallen leaves.

Lightning strikes can cause nitrogen gas to react with oxygen. This can release mineral nitrogen back into the soil – a process called nitrogen fixation.

Some kinds of bacteria help release minerals, such as nitrate, into the soil after feeding on dead leaves. This is called nitrification.

Plants get their nitrogen by absorbing nitrate through their roots.

Oceans contain huge amounts of carbon – about
50 times more than the amount in the atmosphere.

187

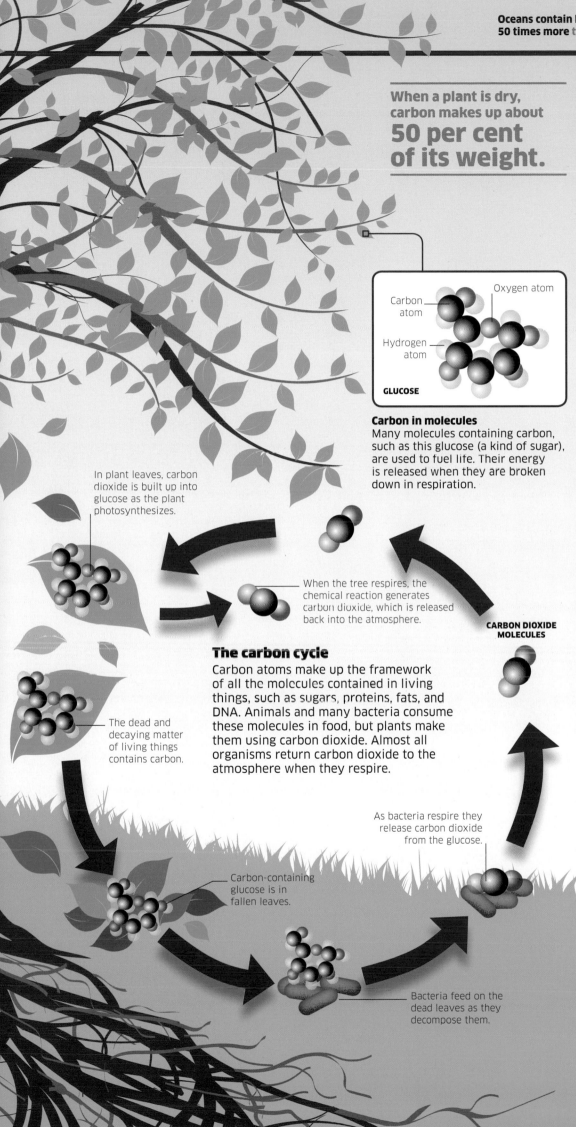

When a plant is dry,
carbon makes up about
50 per cent
of its weight.

Carbon
atom

Oxygen atom

Hydrogen
atom

GLUCOSE

Carbon in molecules
Many molecules containing carbon,
such as this glucose (a kind of sugar),
are used to fuel life. Their energy
is released when they are broken
down in respiration.

In plant leaves, carbon
dioxide is built up into
glucose as the plant
photosynthesizes.

When the tree respires, the
chemical reaction generates
carbon dioxide, which is released
back into the atmosphere.

**CARBON DIOXIDE
MOLECULES**

The carbon cycle
Carbon atoms make up the framework
of all the molecules contained in living
things, such as sugars, proteins, fats, and
DNA. Animals and many bacteria consume
these molecules in food, but plants make
them using carbon dioxide. Almost all
organisms return carbon dioxide to the
atmosphere when they respire.

The dead and
decaying matter
of living things
contains carbon.

As bacteria respire they
release carbon dioxide
from the glucose.

Carbon-containing
glucose is in
fallen leaves.

Bacteria feed on the
dead leaves as they
decompose them.

Recycling water
Water is made of two elements – hydrogen
and oxygen – and travels through earth, sea,
and sky in the global water cycle. This cycle
is dominated by two processes: evaporation
and precipitation. Liquid water in oceans,
lakes, and even on plant leaves evaporates to
form gaseous water vapour. The water vapour
then condenses to form the tiny droplets
inside clouds, before falling back down to
Earth as precipitation: rain, hail, or snow.

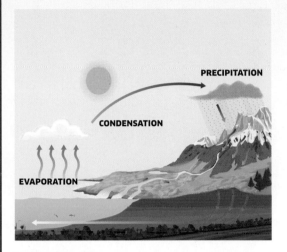

PRECIPITATION

CONDENSATION

EVAPORATION

The water cycle
Recycling of water is driven by the heating
effects of the Sun. At it is warmed, surface water
evaporates into the atmosphere, but cools and
condenses to form rain or snow. Rainfall drains or
runs off to oceans and lakes, to complete the cycle.

The long-term carbon cycle
Carbon atoms can be recycled between living
organisms and the air within days, but other
changes deeper in the earth take place over
millions of years. Lots of carbon gets trapped
within the bodies of dead organisms either
in the ocean or underground – forming fossil
fuels. It is then only released back into the
atmosphere through natural events such as
volcanic eruptions, or when it is burnt by
humans, in forms such as coal (see pp.36–37).

Coal mining
At this mining terminal in Australia, carbon-
containing coal is extracted from the ground.

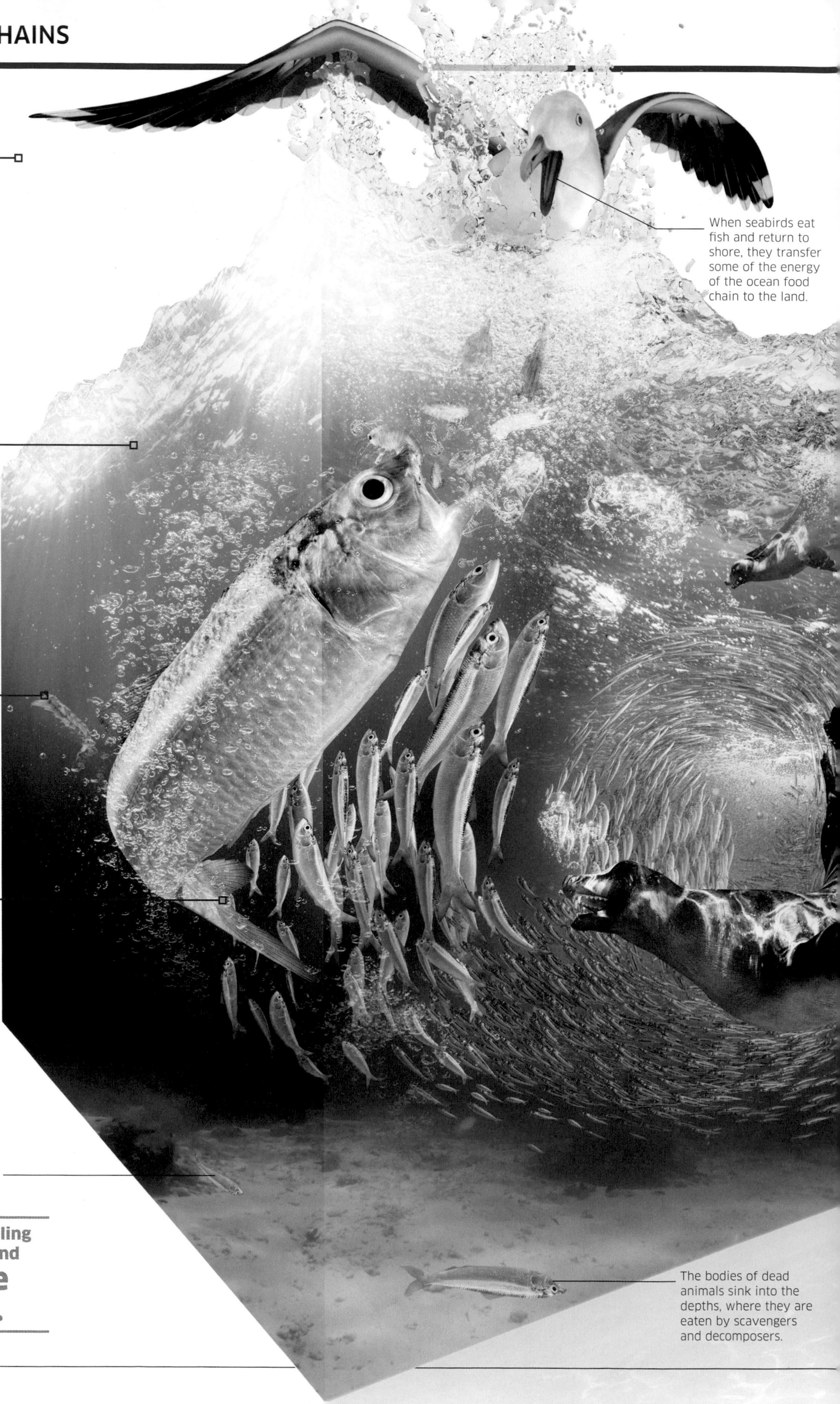

1 Sunlight
When the Sun is shining brightly, a single square metre of ocean surface collects more than a thousand joules of energy every second – enough to power a microwave oven.

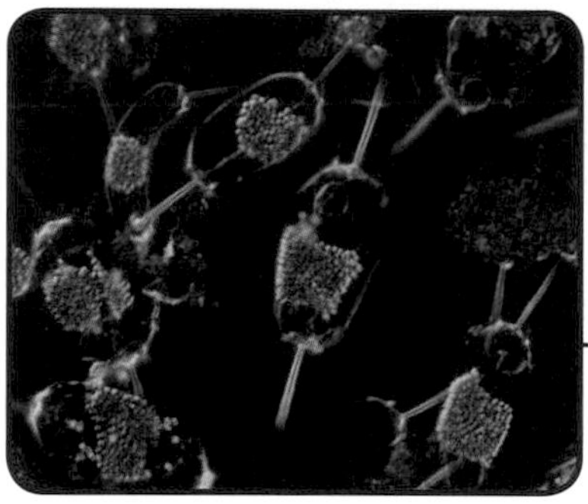

When seabirds eat fish and return to shore, they transfer some of the energy of the ocean food chain to the land.

2 Phytoplankton
Plankton are tiny organisms that float in the water in their billions. They contain algae called phytoplankton that make food by photosynthesis. As they harness their energy directly from the Sun, they are called the producers in a food chain.

3 Zooplankton
Tiny animals, called zooplankton, feed on the phytoplankton. Including a variety of shrimps and fish larvae, these are the primary consumers – animals that eat only algae or plants. They make up the second stage of a food chain.

4 Herring
The Pacific herring is a key link in the ocean food chain. An omnivore that eats both phytoplankton and zooplankton, it is the secondary consumer of the chain, and swims in large shoals that are easily snapped up by bigger predators.

Food waste
In deeper, darker parts of the ocean there is not enough light for photosynthesis, so food chains here often rely on dead organisms falling down through the water.

Photosynthesis by ocean-dwelling phytoplankton generates around
70 per cent of the oxygen in the air.

The bodies of dead animals sink into the depths, where they are eaten by scavengers and decomposers.

Some deep-ocean **food chains start with minerals produced from volcanic vents**, rather than sunlight.

Although most primary consumers must eat large numbers of plants, just **one large tree can support thousands of plant-eaters.**

189

Food chains

Living things rely on one another for nourishment. Energy in a food chain travels from the Sun to plants, then animals, and finally to predators at the very top of the chain.

The Sun provides the ultimate source of energy for life on Earth. Plants and algae change its light energy into chemical energy when they photosynthesize. Vegetarian (herbivorous) animals consume this food and they, in turn, are eaten by meat-eating carnivores. Energy is passed up the chain, and also transfers to scavengers and decomposers (see pp.146–147) when they feed on the dead remains of organisms.

An ocean food chain

Near the surface of the ocean, where bright sunlight strikes the water, billions of microscopic algae photosynthesize to make food. In doing so, they kick-start a food chain that ends with some of the biggest meat-eaters on the planet.

Heat production
The chemical reactions that take place in living organisms generate heat, which escapes into the surrounding water.

Ecological pyramids

The levels of a food chain can be shown stacked up together to make an ecological pyramid. Plants or algae – the producers of food – form the base of the pyramid, with consumers on the higher levels. Each stage of the pyramid can also be shown as the total weight of the organisms on that level – their biomass. Both biomass and, usually, the number of animals decreases towards the top, as energy is lost at each level. Organisms use energy to stay alive and it is given off as waste and heat, leaving less to be passed on.

Only about 10 per cent of the energy, and biomass, in any level passes to the one above.

The amount of biomass decreases at each level.

The number of organisms usually decreases at each higher level of the pyramid.

Primary consumers, such as rabbits, must eat large numbers of plants to get enough energy.

1 kg (2.2 lb)

TOP PREDATOR

10 kg (22 lb)

SECONDARY CONSUMERS

100 kg (220 lb)

PRIMARY CONSUMERS

1,000 kg (2,200 lb)

PRODUCERS

6 Great white shark
Being the food chain's top predator means that little else will prey on an adult great white shark. But, like all other organisms, after death the energy in its body will support decomposers that feed on its corpse.

5 Sealion
Sealions swim hundreds of metres from the shoreline to reach the best fishing grounds. As they hunt herring, the energy in the fish meat passes into the sealion's body. Because their herring prey are also meat-eaters, this makes sealions tertiary consumers.

190 life ○ THREATENED SPECIES

Poaching has driven the **northern white rhinoceros** to the brink of extinction – there are just **two left in the world**.

Threatened species

Human activities, such as habitat destruction and hunting, threaten many species of plants and animals with extinction.

In 1964, the International Union for the Conservation of Nature (IUCN) – the world authority on conservation – started to list endangered species on the Red List. Since then, it has grown to cover thousands of species.

THE RED LIST CRITERIA

Scientists choose a level of threat for each species from among seven categories, depending on the results of surveys and other research. An eighth category includes species that need more study before a decision is made. The numbers of species on the Red List at the end of 2017 are listed below.

- ◎ **Least concern:** 30,385
- ◎ **Near threatened:** 5,445
- ◎ **Vulnerable:** 10,010
- ◎ **Endangered:** 7,507
- ◎ **Critically endangered:** 5,101
- ◎ **Extinct in wild:** 68
- ◎ **Extinct:** 844

Threatened numbers
The Red List has prioritized groups such as amphibians, reptiles, and birds that are thought to be at greatest risk. Most species – especially invertebrates, which make up 97 per cent of all animal species – have not yet been assessed.

Back from the brink
The Mauritius pink pigeon population had fallen to just 10 individuals by 1990. Conservation efforts helped to bring the numbers back up to a possible 380 by 2011.

LEAST CONCERN

Widespread and abundant species facing no current extinction threat: some do well in habitats close to humans and have even been introduced into countries where they are not native.

HUMAN
Homo sapiens
Location: Worldwide
Population: 7.5 billion; increasing

MALLARD
Anas platyrhynchos
Location: Worldwide
Population: 19 million; increasing

CANE TOAD
Rhinella marina
Location: Tropical America, introduced elsewhere
Population: Unknown; increasing

NEAR THREATENED

Species facing challenges that may make them threatened in the near future: a decreasing population size increases risk.

JAGUAR
Panthera onca
Location: Central and South America
Population: 64,000; decreasing

Shaggy, reddish feathers

REDDISH EGRET
Egretta rufescens
Location: Central and South America
Population: Unknown; decreasing

Moist skin

JAPANESE GIANT SALAMANDER
Andrias japonicus
Location: Japan
Population: Unknown; decreasing

PISTACHIO
Pistacia vera
Location: Southwestern Asia
Population: Unknown, decreasing

VULNERABLE

Species that may be spread over a wide range or abundant, but face habitat destruction and hunting.

HUMBOLDT PENGUIN
Spheniscus humboldti
Location: Western South America
Population: 30,000–40,000

Enormous colourful wings

ROTHSCHILD'S BIRDWING
Ornithoptera rothschildi
Location: Western New Guinea
Population: Unknown

GOLDEN HAMSTER
Mesocricetus auratus
Location: Syria, Turkey
Population: Unknown; decreasing

Long, paddle-like snout

AMERICAN PADDLEFISH
Polyodon spathula
Location: Mississippi River Basin
Population: More than 10,000

The **passenger pigeon** was once the **commonest bird in North America**, but hunting drove it to extinction – the last one died in Cincinnati Zoo in 1914.

Conservation projects, such as protecting forest habitats, have increased the number of **giant pandas** in the world – they are **no longer endangered**.

191

ENDANGERED

Species restricted to small areas, with small populations, or both: conservation projects, such as protecting habitats, can help save them from extinction.

WHALE SHARK
Rhincodon typus
Location: Warm oceans worldwide
Population: 27,000–238,000; decreasing

Flat face with forward-facing eyes

CHIMPANZEE
Pan troglodytes
Location: Central Africa
Population: 173,000–300,000; decreasing

FIJIAN BANDED IGUANA
Brachylophus bulabula
Location: Fiji
Population: More than 6,000; decreasing

GURNEY'S PITTA
Hydrornis gurneyi
Location: Myanmar, Thailand
Population: 10,000–17,200; decreasing

Yellow and black under parts on males

CRITICALLY ENDANGERED

Species in greatest danger: some have not been seen in the wild for so long that they may already be extinct; others have plummeted in numbers.

YANGTZE RIVER DOLPHIN
Lipotes vexillifer
Location: Yangtze River
Population: Last seen 2002; possibly extinct

COMMON SKATE
Dipturus batis
Location: Northeastern Atlantic
Population: Unknown; decreasing

SPIX'S MACAW
Cyanopsitta spixii
Location: Brazil
Population: Last seen 2016; possibly extinct in wild

Blue plumage

CHINESE ALLIGATOR
Alligator sinensis
Location: China
Population: Possibly fewer than 150 in wild

Thick armoured skin

EXTINCT IN WILD

Species that survive in captivity or in cultivation: a few, such as Père David's deer, have been reintroduced to wild habitats from captive populations.

GUAM KINGFISHER
Todiramphus cinnamominus
Last wild record: Guam, 1986
Population: 124 in captivity

BLACK SOFTSHELL TURTLE
Nilssonia nigricans
Last wild record: Bangladesh, 2002
Population: 700 in artificial pond

PÈRE DAVID'S DEER
Elaphurus davidianus
Last wild record: China, 1800 years ago
Population: Large captive herds; reintroduced to wild

Long, backwards-pointing antlers in males

WOOD'S CYCAD
Encephalartos woodii
Last wild record: South Africa, 1916
Population: A handful of clones of one plant in botanic gardens

EXTINCT

Species no longer found alive in the wild, even after extensive surveys, nor known to exist in captivity or cultivation: under these circumstances, it is assumed that the last individual has died.

GOLDEN TOAD
Incilius periglenes
Last wild record: Costa Rica, 1989
Population: Declared extinct 2004

CAROLINA PARAKEET
Conuropsis carolinensis
Last wild record: USA, 1904
Population: Last parakeet died in zoo, 1918

THYLACINE
Thylacinus cynocephalus
Last wild record: Tasmania, 1930
Population: Last thylacine died in zoo, 1936

ST HELENA GIANT EARWIG
Labidura herculeana
Last wild record: St Helena, 1967
Population: Declared extinct 2001

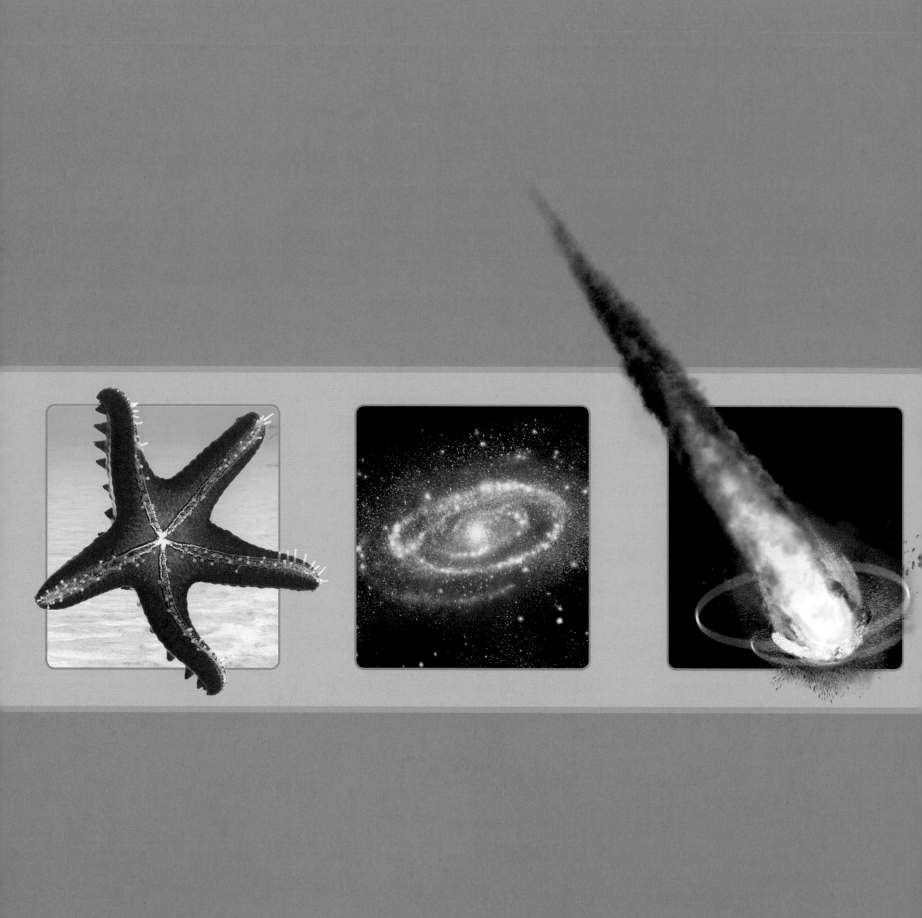

REFERENCE

The scope of science stretches far and wide. Scientists study the vast expanse of the Universe and everything within it – including the diversity of life and how it evolved. Careful observation, measurements, and experiments help scientists understand the world.

There are more **stars in the Universe** than there are **grains of sand on all the beaches on Earth.**

Scale of the Universe

The difference in size between the smallest and biggest things in the Universe is unimaginably vast – from subatomic particles to galaxies.

No-one knows how big the Universe is, but it has been expanding since it formed in the Big Bang 13.7 billion years ago. The distances are so great that cosmologists measure them in terms of light years – the distance light moves in space in a year, which is equal to 9.5 trillion kilometres (6 trillion miles) – and parts of the Universe are billions of light years apart.

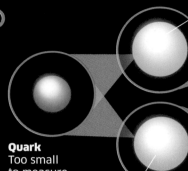

Proton
A particle in an atomic nucleus that carries a positive charge.

Quark
Too small to measure, different kinds of quarks are the subatomic particles that make up protons and neutrons.

Neutron
A particle in an atomic nucleus that carries no charge.

Carbon atom
With six electrons orbiting a nucleus of six protons and six neutrons, a carbon atom is less than a billionth of a metre across.

Limestone rock
Solid mixtures of billions of tiny fossilized shells and mineral fragments make up limestone rock, containing calcium carbonate – a compound that has atoms of calcium, carbon, and oxygen.

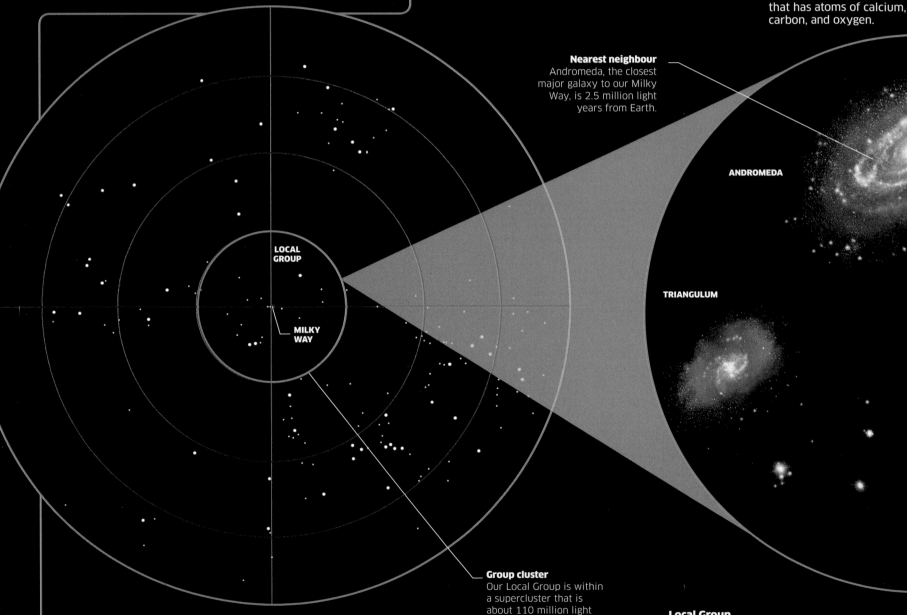

Nearest neighbour
Andromeda, the closest major galaxy to our Milky Way, is 2.5 million light years from Earth.

ANDROMEDA

TRIANGULUM

LOCAL GROUP

MILKY WAY

Group cluster
Our Local Group is within a supercluster that is about 110 million light years in diameter.

Supercluster
Clusters of galaxies span a region of space ten times bigger than the Local Group. Such a supercluster can contain tens of thousands of galaxies. Our Milky Way is within the Virgo Supercluster. Scientists think there are about 10 million superclusters in the observable Universe.

Local Group
The Milky Way is part of a so-called Local Group of about 50 galaxies that stretch across 10 million light years of space, or 100 times the diameter of our Milky Way. Galaxies are millions of times further apart than the stars that are in each one. Andromeda is the biggest galaxy in our Local Group. Most others are much smaller.

Travelling at the **speed of light**, it would take **100,000 years** to cross the Milky Way.

The **Sun** accounts for **99.8 per cent** of the mass of our Solar System.

Thousands of **exoplanets** have been discovered **outside our Solar System** since the first one was identified in 1995.

195

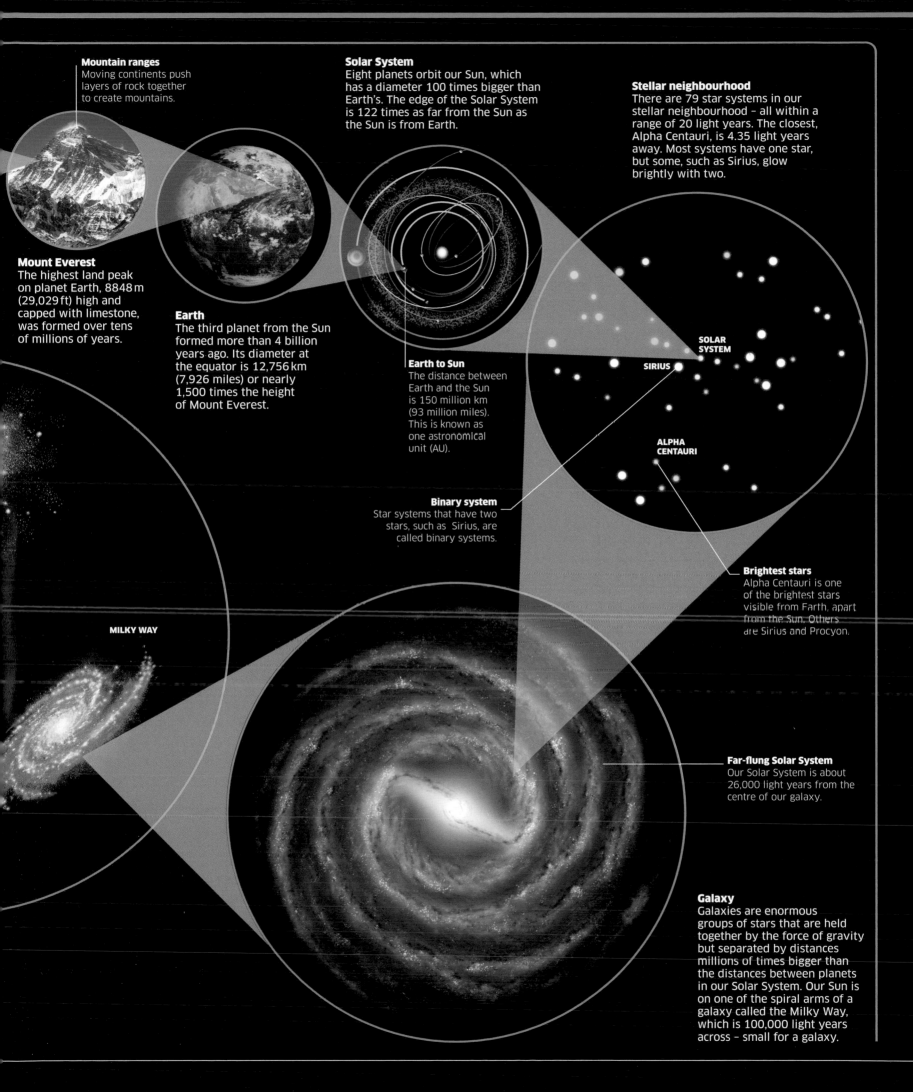

Mountain ranges
Moving continents push layers of rock together to create mountains.

Mount Everest
The highest land peak on planet Earth, 8848 m (29,029 ft) high and capped with limestone, was formed over tens of millions of years.

Earth
The third planet from the Sun formed more than 4 billion years ago. Its diameter at the equator is 12,756 km (7,926 miles) or nearly 1,500 times the height of Mount Everest.

Solar System
Eight planets orbit our Sun, which has a diameter 100 times bigger than Earth's. The edge of the Solar System is 122 times as far from the Sun as the Sun is from Earth.

Earth to Sun
The distance between Earth and the Sun is 150 million km (93 million miles). This is known as one astronomical unit (AU).

Stellar neighbourhood
There are 79 star systems in our stellar neighbourhood – all within a range of 20 light years. The closest, Alpha Centauri, is 4.35 light years away. Most systems have one star, but some, such as Sirius, glow brightly with two.

SOLAR SYSTEM

SIRIUS

ALPHA CENTAURI

Binary system
Star systems that have two stars, such as Sirius, are called binary systems.

Brightest stars
Alpha Centauri is one of the brightest stars visible from Earth, apart from the Sun. Others are Sirius and Procyon.

MILKY WAY

Far-flung Solar System
Our Solar System is about 26,000 light years from the centre of our galaxy.

Galaxy
Galaxies are enormous groups of stars that are held together by the force of gravity but separated by distances millions of times bigger than the distances between planets in our Solar System. Our Sun is on one of the spiral arms of a galaxy called the Milky Way, which is 100,000 light years across – small for a galaxy.

Units of measurement

Scientists measure quantities – such as length, mass, or time – using numbers, so that their sizes can be compared. For each kind of quantity, these measurements must be in units that mean the same thing wherever in the world the measurements are made.

SI units
The abbreviation "SI" stands for *Système International*. It is a standard system of metric units that has been adopted by scientists all over the world so that all their measurements are done in the same way.

Base quantities

Just seven quantities give the most basic information about everything around us. Each is measured in SI units and uses a symbol as an abbreviation. The SI system is metric, meaning that smaller and larger units are obtained by dividing or multiplying by 10, 100, 1000, etc. Centimetres, for instance, are 100 times smaller than a metre, but kilometres are 1,000 times bigger.

LENGTH
SI unit: metre (m)

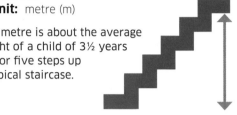

One metre is about the average height of a child of 3½ years old, or five steps up a typical staircase.

- A millionth of a metre (1 micrometre) = the length of a bacterium.
- A thousandth of a metre (1 millimetre) = the diameter of a pinhead.
- 1,000 metres (1 kilometre) = the average distance an adult walks in 12 minutes.

MASS
SI unit: kilogram (kg)

One kilogram is the mass of one litre of water, or about the mass of an average-sized pineapple.

- A thousand trillionth of a kilogram (1 picogram) = the mass of a bacterium.
- A thousandth of a kilogram (1 gram) = the mass of a paper clip.
- 1,000 kilograms (1 metric tonne) = the average mass of an adult walrus.

TIME
SI unit: second (s)

One second is the time it takes to swallow a mouthful of food, or to write a single-digit number.

- A thousandth of a second (1 millisecond) = the time taken by the brain to fire a nerve impulse.
- A tenth of a second (1 decisecond) = a blink of an eye.
- 1 billion seconds (1 gigasecond) = 32 years.

TEMPERATURE
SI unit: kelvin (K)

Just one degree rise in temperature can make you feel hot and feverish.

Temperature scales
An everyday temperature scale most often uses degrees Celsius (°C), and is divided into 100 units between the boiling and freezing points of water at sea level. Kelvin measures all the way down to absolute zero, where heat energy does not exist.

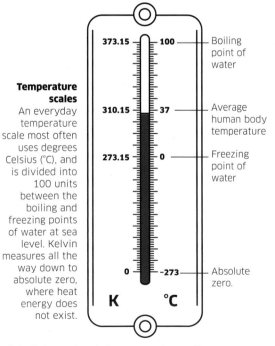

K	°C	
373.15	100	Boiling point of water
310.15	37	Average human body temperature
273.15	0	Freezing point of water
0	−273	Absolute zero.

- 0 kelvin = absolute zero, when all objects and their particles are still.
- 1 kelvin = the coldest known object in the Universe, the Boomerang Nebula.
- 1,000 kelvin = the temperature in a charcoal fire.

ELECTRICAL CURRENT
SI unit: ampere (A)

One ampere is about the current running through a 100W light bulb.

- A thousandth of an ampere (1 milliampere) = the current in a portable hearing aid.
- 100,000 amperes = the current in the biggest lightning strikes.
- 10 thousand billion amperes = the current in the spiral arms of the Milky Way.

LIGHT INTENSITY
SI unit: candela (cd)

One candela is the light intensity given off by a candle flame.

- A millionth of a candela = the lowest light intensity perceived by human vision.
- A thousandth of a candela = a typical night sky away from city lights.
- 1 billion candelas = the intensity of the Sun when viewed from the Earth.

AMOUNT OF A SUBSTANCE
SI unit: mole (mol)

One mole is a fixed number of atoms, molecules, or other particles. Because substances all have different atomic structures, one mole of one substance may be very different to that of another.

A mole of gold atoms is in about six gold coins.

A mole of sugar molecules fills about two small cups.

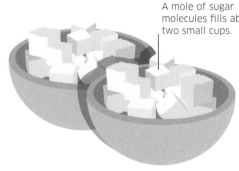

- A tenth of a mole of iron atoms = the amount of iron in the human body.
- 1,000 moles of carbon atoms = the amount of carbon in the human body.
- 10 million trillion moles of oxygen molecules = the amount of oxygen in the Earth's atmosphere.

Derived quantities

Other kinds of quantities are also useful in science, but these are calculated from base quantities using scientific equations. For instance, we combine SI measurements of mass, distance, and time to work out an SI measurement for force. This means that force is said to be a derived quantity.

FORCE
SI unit: newton (N)

One newton is about the force of gravity on a single apple.

$$\text{Force in newtons} = \frac{\text{Mass in kilograms x distance in metres}}{\text{Time in seconds}^2}$$

- A 10 billionth of a newton = the force needed to break six chemical bonds in a molecule.
- 10 newtons = the weight of an object with mass of 1 kilogram.

PRESSURE
SI unit: pascal (Pa)

One pascal is about the pressure of a bank note resting on a flat surface.

$$\text{Pressure in pascals} = \frac{\text{Force in newtons}}{\text{Area in metres}^2}$$

- A 10 thousand trillionth of a pascal = the lowest pressure recorded in outer space.
- 1 million pascals (1 megapascal) = the pressure of a human bite.

ENERGY
SI unit: joule (J)

One joule is about the energy needed to lift a medium-sized tomato a height of one metre.

$$\text{Energy in joules} = \text{Force in newtons X distance in metres}$$

- A millionth of a joule (1 microjoule) = the energy of motion in six flying mosquitoes.
- 1,000 joules (1 kilojoule) = the maximum energy from the Sun reaching 1 square metre of Earth's surface each second.

FREQUENCY
SI unit: hertz (Hz)

One hertz is about the frequency of a human heartbeat: one beat per second.

$$\text{Frequency in hertz} = \frac{\text{Number of cycles}}{\text{Time in seconds}}$$

- 100 hertz = the frequency of an engine cycle in a car running at maximum speed.
- 10,000 hertz = the frequency of radio waves.

POWER
SI unit: watt (W)

One watt is about the power used by a single Christmas tree light.

$$\text{Power in watts} = \frac{\text{Energy in joules}}{\text{Time in seconds}}$$

- A millionth of a watt (1 microwatt) = the power used by a wristwatch.
- 1 billion watts (1 gigawatt) = the power used by a hydroelectric generating station.

POTENTIAL DIFFERENCE
SI unit: volt (V)

Voltage is a measure of the difference in electrical energy between two points – the force needed to make electricity move. One volt is about the voltage in a lemon battery cell.

$$\text{Potential difference in volts} = \frac{\text{Power in watts}}{\text{Current in amperes}}$$

- 100 volts = the mains voltage in the USA.

ELECTRICAL CHARGE AND RESISTANCE
SI unit: Charge – coulomb (C)
Resistance – ohm (Ω)

The measurements relating to electricity are all interlinked. Charge is a measure of how positive or negative particles are, and can be worked out from the current and the time. Resistance is a measure of the difficulty a current has in flowing, and can be worked out from the voltage and the current.

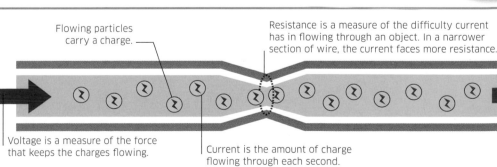

Flowing particles carry a charge.

Resistance is a measure of the difficulty current has in flowing through an object. In a narrower section of wire, the current faces more resistance.

Voltage is a measure of the force that keeps the charges flowing.

Current is the amount of charge flowing through each second.

$$\text{Resistance in ohms} = \frac{\text{Potential difference in volts}}{\text{Current in amperes}}$$

$$\text{Charge in coulombs} = \text{Current in amperes X time in seconds}$$

Classifying life

Scientists have described more than a million and a half different species of living things. They classify them into groups based on how they are related.

There are lots of ways of classifying organisms. Insects, birds, and bats could be grouped as flying animals, and plants could be grouped by how we use them. But neither of these systems shows natural relationships. Biological classification works by grouping related species. Bats, for instance, have closer links to monkeys than they do to birds, because they are both furry mammals that have evolved from the same mammal ancestors.

Single-celled organisms are the most common form of life in some kingdoms, including archaea, bacteria, and protozoans.

Seven kingdoms of organisms

> Archaea Bacteria Protozoans

Over 30 phyla of animals, including...

> Flatworms Annelids Molluscs

12 classes of chordates, including...

> Sea squirts Jawless fishes Cartilaginous fishes Lobe-finned fishes Ray-finned fishes

29 orders of mammals, including...

> Monotremes Marsupials Elephants Sloths and anteaters Primates Rodents Rabbits, hares, and pikas

15 families of primates, including...

> Dwarf and mouse lemurs True lemurs Sifakas and relatives Bushbabies Aye-aye Lorises and relatives Tarsiers

Scientific names

Every species has a two-part scientific name using Latin words that are internationally recognised in science. The first part identifies its genus group, the second its species. Lions and tigers belong to the *Panthera* genus of big cats, but have different species names.

Panthera leo
Lion

Panthera tigris
Tiger

MAMMALS TURTLES LIZARDS AND SNAKES CROCODILES BIRDS

Fossils and DNA show that birds are most closely related to crocodiles.

Classifying birds

Modern classification aims to show how organisms are linked by evolution. Birds and reptiles are traditionally in two separate classes. But birds evolved from reptile ancestors (see pp.136–137), so many scientists think they should be a sub-group of the reptiles.

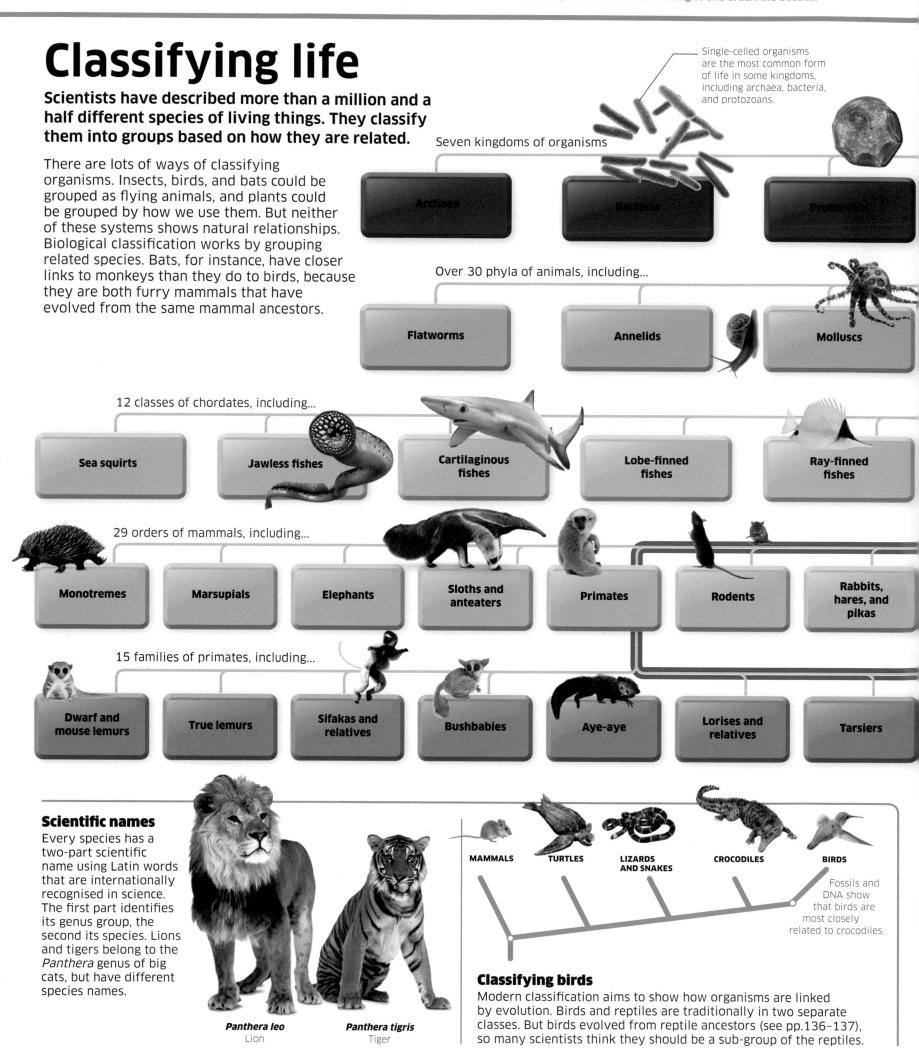

90 per cent of plant species are flowering plants. The remainder are spore-producing plants, such as mosses.

Most species on Earth still await discovery. It's possible that **less than 20 per cent** have been classified so far.

199

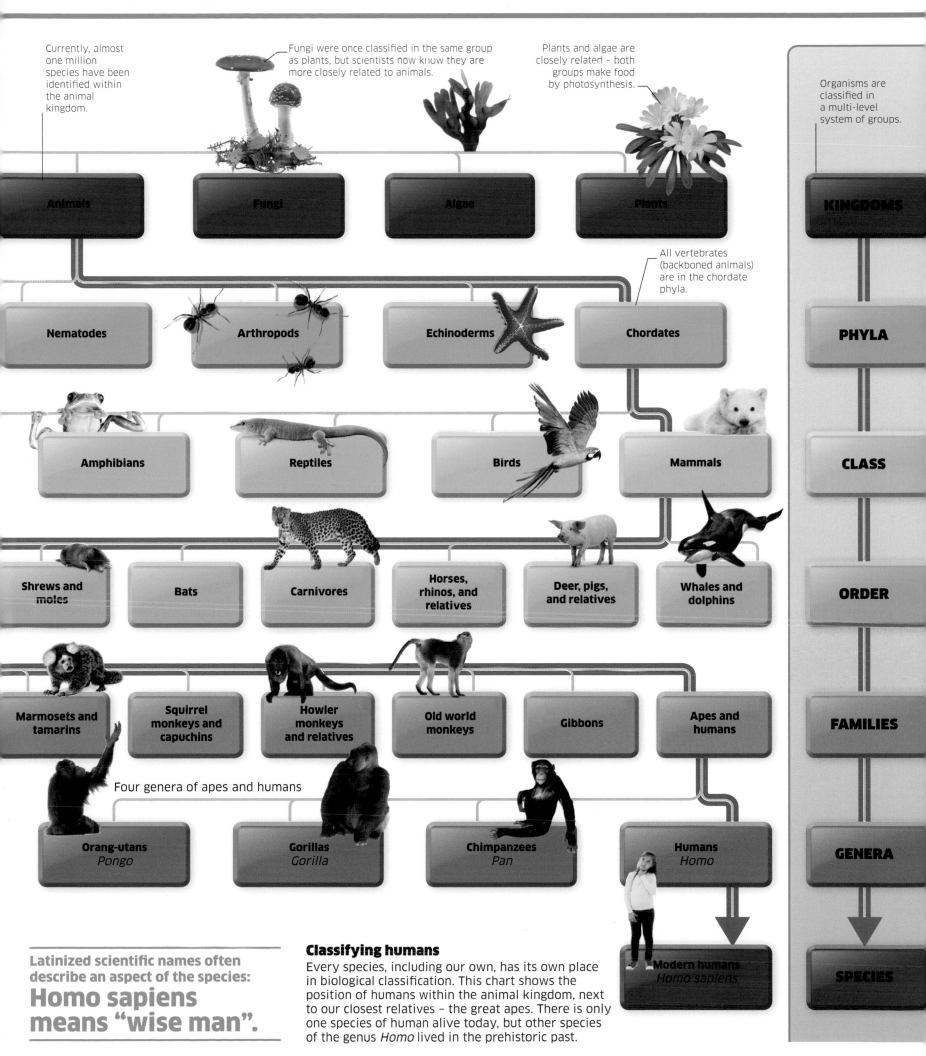

Currently, almost one million species have been identified within the animal kingdom.

Fungi were once classified in the same group as plants, but scientists now know they are more closely related to animals.

Plants and algae are closely related – both groups make food by photosynthesis.

Organisms are classified in a multi-level system of groups.

Animals

Fungi

Algae

Plants

KINGDOMS

All vertebrates (backboned animals) are in the chordate phyla.

Nematodes

Arthropods

Echinoderms

Chordates

PHYLA

Amphibians

Reptiles

Birds

Mammals

CLASS

Shrews and moles

Bats

Carnivores

Horses, rhinos, and relatives

Deer, pigs, and relatives

Whales and dolphins

ORDER

Marmosets and tamarins

Squirrel monkeys and capuchins

Howler monkeys and relatives

Old world monkeys

Gibbons

Apes and humans

FAMILIES

Four genera of apes and humans

Orang-utans
Pongo

Gorillas
Gorilla

Chimpanzees
Pan

Humans
Homo

GENERA

Modern humans
Homo sapiens

SPECIES

Latinized scientific names often describe an aspect of the species:

Homo sapiens means "wise man".

Classifying humans

Every species, including our own, has its own place in biological classification. This chart shows the position of humans within the animal kingdom, next to our closest relatives – the great apes. There is only one species of human alive today, but other species of the genus *Homo* lived in the prehistoric past.

Modern humans appeared just over a quarter of a million years ago – a fraction of the timeline for life on Earth.

Elasmosaurus
The largest reptiles in the oceans included predatory, long-necked plesiosaurs.

Comet impact
A decline in sea levels followed by a comet strike brought an end to the dinosaurs.

65 MYA

PALEOGENE

Paleogene Period
Mammals replaced dinosaurs as the dominant large animals, and many of them grazed on a new kind of plant that grew on open land: grass.

Tyrannosaurus
Dinosaurs, including formidable predators, became the largest ever land-living animals.

CRETACEOUS-PALEOGENE EXTINCTION

145 MYA

CRETACEOUS

Cretaceous Period
Flowering plants grew on lands ruled by reptiles until an asteroid impact wiped out half of all species, including all the dinosaurs and other giant reptiles – paving the way for mammals.

Pteranodon
Winged reptiles included the largest animals ever to fly.

Archaeopteryx
The first birds – evolving from dinosaur ancestors – took to the air.

Sternopterygius
Dolphin-shaped ichthyosaurs swimming in the Jurassic seas preyed on fish.

305 MYA

Timeline of life

Half a billion years ago the only living things were small and simple. Over time, evolution has produced a spectacular world of plants and animals.

The timeline of life is divided into periods that were dominated by particular kinds of organisms – such as invertebrates, fishes, or reptiles. As the surface of Earth changed, some organisms succeeded, while others died away. Continents shifted, seas rose and fell, and luxuriant forests turned into parched deserts and back again. Catastrophes, such as asteroid strikes or ice ages, even drove some major groups to extinction. Such events all made their mark on living things. However, throughout the history of Earth, life went on as descendants after descendants eventually led to the natural world we know today.

Carboniferous coal forest
Trees grew tall in the warm Carboniferous swamps. Their remains left the coal deposits we know today.

Eryops
Amphibian descendants of fishes became large, backboned animals.

Meganeura
Insects such as this crow-sized dragonfly were the first animals to fly.

359 MYA

CARBONIFEROUS

Carboniferous Period
Warm, rich swamp forests provided the perfect habitat for giant amphibians and insects, while the first hard-shelled eggs were laid by the earliest reptiles.

CAMBRIAN EXPLOSION

541 MYA

Cambrian Period
Once multi-celled animals appeared they evolved into an explosion of many different body forms – producing the first representatives of all the major groups alive today.

Halkoulchthys
With two eyes and simple fins, this small animal was related to our backboned ancestors.

Anomalocaris
Early animals, such as this aquatic invertebrate, were unlike any alive today.

488 MYA

ORDOVICIA

The earliest evidence of **single-celled fossil life** exists in rocks that are **3.5 billion years old**.

Billions of years ago, algae produced much of the oxygen in the air today.

201

Megacerops
Tiny ancestors evolved into large mammals that replaced the dinosaurs.

23 MYA

NEOGENE

Neogene Period
Many familiar groups of mammals, such as rodents, primates, antelopes, and cats, evolved in the Neogene, while flying birds diversified in the skies.

Thylacosmilus
Predatory mammals, including sabre-toothed cats, hunted grazers on the open grassland.

2.6 MYA

QUATERNARY

Quaternary Period
Mammals and birds survived Quaternary ice ages, but one species (humans) drove many others to extinction by hunting and habitat destruction in a mass extinction of modern times.

201 MYA

Red lines indicate mass extinctions.

TRIASSIC-JURASSIC EXTINCTION

Jurassic Period
The peak of the Age of Dinosaurs saw the evolution of giant reptiles on land and in the seas, which included the largest land animals of all time.

JURASSIC

Asteroid impact
A space rock colliding with Earth wiped out a quarter of all species.

Eoraptor
The first dinosaurs, some two-legged, evolved from small reptiles that survived the Permian-Triassic extinction.

Periods of time
Earth's prehistory is marked by a timescale divided up into geological periods. Each period represents a length of time that has left its mark in rocks by fossils and other evidence.

299 MYA

PERMIAN-TRIASSIC EXTINCTION

Permian Period
Continents dried up to favour scaly reptiles over amphibians. The period ended with violent eruptions causing the biggest of all mass extinctions.

PERMIAN

Dimetrodon
As moist-skinned amphibians declined in a drier world, reptiles such as this carnivore took over.

Triassic Period
New kinds of forests containing conifers and cycads were inhabited by the first dinosaurs until a possible asteroid impact brought about mass extinction.

TRIASSIC

251 MYA

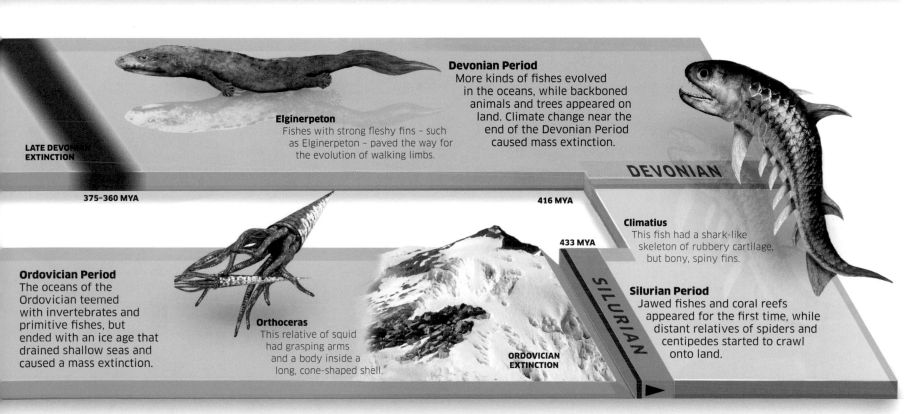

Devonian Period
More kinds of fishes evolved in the oceans, while backboned animals and trees appeared on land. Climate change near the end of the Devonian Period caused mass extinction.

Elginerpeton
Fishes with strong fleshy fins – such as Elginerpeton – paved the way for the evolution of walking limbs.

LATE DEVONIAN EXTINCTION

DEVONIAN

375–360 MYA

416 MYA

Climatius
This fish had a shark-like skeleton of rubbery cartilage, but bony, spiny fins.

433 MYA

Ordovician Period
The oceans of the Ordovician teemed with invertebrates and primitive fishes, but ended with an ice age that drained shallow seas and caused a mass extinction.

Orthoceras
This relative of squid had grasping arms and a body inside a long, cone-shaped shell.

ORDOVICIAN EXTINCTION

SILURIAN

Silurian Period
Jawed fishes and coral reefs appeared for the first time, while distant relatives of spiders and centipedes started to crawl onto land.

Glossary

ACID
A substance with a pH lower than 7.

ALGAE
Plant-like organisms that can make food using energy from sunlight.

ALKALI
See Base

ALLOY
A mixture of two or more metals, or of a metal and a non-metal.

ANALOGUE
Relating to signals or information represented by a continuously varying value, such as a wave.

ATMOSPHERE
The layer of breathable gases, such as oxygen and nitrogen, that surrounds Earth.

ATOM
The smallest unit of an element.

BACTERIA
Microscopic organisms with a simple, single-celled form.

BASE
A substance with pH higher than 7. Bases that are soluble in water are called alkalis. Also: one of the four chemicals that make up the "rungs" of a DNA double helix.

BIOLOGY
The science of living things.

BUOYANCY
The tendency of a solid to float or sink in liquids.

CARBOHYDRATE
An energy-rich substance, such as sugar or starch.

CATALYST
A substance that makes a chemical reaction occur much more rapidly, but is not changed by the reaction.

CELL
The smallest unit of life.

CHEMICAL BOND
An attraction between particles, such as atoms or ions.

CHEMICAL REACTION
A process that changes substances into new substances by breaking and making chemical bonds.

CHEMISTRY
The science of matter and elements.

CHROMOSOME
A threadlike structure, found in the nucleus of cells, that is made up of coiled strands of DNA. Humans have 46 chromosomes per body cell.

CLIMATE
The most common weather conditions in an area over a long period of time.

COMBUSTION
A chemical reaction in which a substance reacts with oxygen, releasing heat and flames.

COMPOUND
A chemical substance in which two or more elements have bonded together.

CONCENTRATION
The amount of one substance mixed in a known volume of the other.

CONDENSATION
A process whereby a gas changes into a liquid.

CONDUCTOR
A substance through which heat or electric current flows easily.

COVALENT BOND
A type of chemical bond in a molecule where atoms share one or more electrons.

DNA
A material found in the cells of all organisms that carries instructions for how a living thing will look and function.

DRAG
The resistance force formed when an object pushes through a fluid, such as air or water.

ECOSYSTEM
A community of organisms and the non-living environment around them.

ELECTRIC CHARGE
How positive or negative a particle is.

ELECTRON
One of the tiny particles inside an atom. It has a negative electric charge.

ELEMENT
A simple substance made of atoms that are all the same kind.

ENERGY
What enables work to be done. Energy exists in many different forms and cannot be created or destroyed, only transferred.

ENZYME
A substance produced in living organisms that acts as a catalyst and speeds up chemical reactions.

EROSION
A process by which Earth's surface rocks and soil are worn away by wind, water, or ice.

EVAPORATION
A process by which a liquid changes into a gas.

EVOLUTION
The process by which Earth's species gradually change over long periods of time, such as millions of years, to produce new species.

EXCRETION
The process by which living organisms expel or get rid of waste produced by cells of the body.

FERTILIZATION
The joining of male and female sex cells so they develop into new life.

FISSION
A splitting apart; nuclear fission is the splitting of the nucleus of an atom.

FOSSIL
The preserved remains or impressions of life from an earlier time.

FOSSIL FUEL
A substance formed from the remains of ancient organisms that burns easily to release energy.

FRICTION
The dragging force that occurs when one object moves over another.

FUSION
A joining together; nuclear fusion is the joining of two atomic nuclei.

GAS
A state of matter that flows to fill a container, and can be compressed.

GENE
One of the tiny units carried on DNA that determine what a living thing looks like and how it functions.

GLUCOSE
A simple carbohydrate, or sugar, made by photosynthesis and then used by cells as a source of energy.

GRAVITY
The force that attracts one object to another and prevents things on Earth from floating off into space.

HABITAT
The area where an animal naturally makes its home.

INHERITANCE
The range of natural characteristics passed on to offspring by parents.

INSULATOR
A material that stops heat moving from a warm object to a colder one.

ION
An atom that has lost or gained one or more electrons and as a result has either a positive or negative electric charge.

IONIC BOND
A type of chemical bond where one or more electrons are passed from one atom to another, creating two ions of opposite charge that attract each other.

ISOTOPE
One of two or more atoms of a chemical element that have different numbers of neutrons compared to other atoms of the element.

LIFT
The upward force produced by an aircraft's wings that keeps it airborne.

LIQUID
A state of matter that flows and takes the shape of a container, and cannot be compressed.

MAGMA
Hot, liquid rock that is found beneath Earth's surface.

MAGNET
An object that has a magnetic field and attracts or repels other magnetic objects.

MASS
A measure of the amount of matter in an object.

MATERIAL
A chemical substance out of which things can be made.

METAL
Any of many elements that are usually shiny solids and good conductors of electricity.

MICROORGANISM
A tiny organism which can only be seen with the aid of a microscope. Also known as a microbe.

MINERAL
A solid, non-living material occurring naturally in Earth made up of a particular kind of chemical compound.

MOLECULE
A particle formed by two or more atoms joined by covalent bonds.

MONOMER
A molecule that can be bonded to other similar molecules to form a polymer.

NERVE
A fibre that carries electrical messages (nerve impulses) from one part of the body to another.

NEUTRON
One of the tiny particles in an atom. It has no electric charge.

NUCLEUS
The control centre inside the cells of most living organisms. It contains genetic material, in the form of DNA. Also: the central part of an atom, made of protons and neutrons.

NUTRIENT
A substance essential for life to exist and grow.

ORBIT
The path taken by an object, for example a planet, that is circling around another.

ORGAN
A group of tissues that makes up a part of the body with a special function. Important organs include the heart, lungs, liver, and kidneys.

ORGANISM
A living thing.

PARTICLE
A tiny speck of matter.

PHOTOSYNTHESIS
The process by which green plants use the Sun's energy to make carbohydrates from carbon dioxide and water.

PHYSICS
The science of matter, energy, forces, and motion.

PIGMENT
A chemical substance that colours an object.

POLLEN
Tiny grains produced by flowers, which contain the male cells needed to fertilize eggs.

POLYMER
A long, chain-like molecule made up of smaller molecules connected together.

PRESSURE
The amount of force that is applied to a surface per unit of area.

PRODUCT
A substance produced by a chemical reaction.

PROTEIN
A type of complex chemical found in all living things, used as enzymes and in muscles.

PROTON
One of the tiny particles inside an atom. It has a positive electric charge.

RADIATION
Waves of energy that travel through space. Radiation includes visible light, heat, X-rays, and radio waves. Nuclear radiation includes subatomic particles and fragments of atoms.

RADIOACTIVE
Describing a material that is unstable because the nuclei of its atoms split to release nuclear radiation.

REACTANT
A substance that is changed in a chemical reaction.

REACTIVE
A substance that is likely to become involved in a chemical reaction.

RESPIRATION
The process occurring in all living cells that releases energy from glucose to power life.

ROOM TEMPERATURE
A standard scientific term for comfortable conditions (for humans), usually a temperature of around 20°C (68°F).

SEX ORGANS
The organs of an organism that allow it to reproduce. They usually produce sex cells: sperm in males, and eggs in females.

SOLID
A state of matter in which an element's atoms are joined together in a rigid structure.

SOLUTE
A substance that becomes dissolved in another.

SOLVENT
A substance that can have other substances dissolved in it.

SYNTHETIC
Man-made chemical.

TISSUE
A group of similar cells that carry out the same function, such as muscle tissue, which can contract.

TOXIC
Causing harm, such as a poison.

ULTRASOUND
Sound with a frequency above that which the human ear can detect.

ULTRAVIOLET
A type of electromagnetic radiation with a wavelength shorter than visible light.

UNIVERSE
The whole of Space and everything it contains.

VOLCANO
An opening in Earth's crust that provides an outlet for magma when it rises to the surface.

WAVE
Vibration that transfers energy from place to place, without transferring the matter that it is flowing through.

WAVELENGTH
The distance between wave crests, usually when referring to sound waves or electromagnetic waves.

WEIGHT
The force applied to a mass by gravity.

Index

Acknowledgments

The publisher would like to thank the following people for their assistance in the preparation of this book:

Ben Morgan for editorial and scientific advice; Ann Baggaley, Jessica Cawthra, Sarah Edwards, and Laura Sandford for editorial assistance; Caroline Stamps for proofreading; Helen Peters for the index; Simon Mumford for maps; Phil Gamble, KJA-Artists.com, and Simon Tegg for illustrations; avogadro.cc/cite and www.povray.org for 3D molecular modelling and rendering software.

DK Delhi:
Manjari Rathi Hooda: Head, Digital Operations
Nain Singh Rawat: Audio Video Production Manager
Mahipal Singh, Alok Singh: 3D Artists

Smithsonian Enterprises:
Kealy E. Gordon: Product Development Manager
Ellen Nanney: Licensing Manager
Brigid Ferraro: Vice President, Education and Consumer Products
Carol LeBlanc: Senior Vice President, Education and Consumer Products

Curator for the Smithsonian:
Dr. F. Robert van der Linden, Curator of Air Transportation and Special Purpose Aircraft, National Air and Space Museum, Smithsonian

The Smithsonian name and logo are registered trademarks of the Smithsonian Institution.

The publisher would like to thank the following for their kind permission to reproduce photographs:

(Key: a-above; b-below/bottom; c-centre; f-far; l-left; r-right; t-top)

2 123RF.com: Konstantin Shaklein (tl). **3 Dorling Kindersley:** Clive Streeter / The Science Museum, London (cb). **TurboSquid:** Witalk73 (cra). **6 TurboSquid:** 3d_molier International (c). **10 123RF.com:** scanrail (ca). **Dorling Kindersley:** Ruth Jenkinson / Holts Gems (cra). **11 Dorling Kindersley:** Stephen Oliver (cb). **Dreamstime.com:** Dirk Ercken / Kikkerdirk (cla); Grafner (ca); Ron Sumners / Sumnersgraphicsinc (cl); Kellyrichardsonfl (cb/leaves); Heike Falkenberg / Dslrpix (br). **13 Dreamstime.com:** Wisconsinart (cra). **Science Photo Library:** Dennis Kunkel Microscopy (cra/Cellulose). **15 123RF.com:** molekuul (crb). **19 Dreamstime.com:** Fireflyphoto (cr). **20 Gary Greenberg, PhD / www.sandgrains.com:** (clb). **21 Alamy Stock Photo:** Jim Snyders (tl). **National Geographic Creative:** David Liittschwager (cb). **22 Alamy Stock Photo:** Evan Sharboneau (bl). **Dreamstime.com:** Photographerlondon (cra). **Getty Images:** Wu Swee Ong (bc). **23 Dorling Kindersley:** Ruth Jenkinson / Holts Gems (ca). **Dreamstime.com:** Ali Ender Birer / Enderbirer (cl). **Getty Images:** Alain Bachellier (bl). **24 Alamy Stock Photo:** Björn Wylezich (fclb). **Dorling Kindersley:** Natural History Museum, London (bc). **Getty Images:** De Agostini / A. Rizzi (clb). **National Museum of Natural History, Smithsonian Institution:** (cb). **24-25 Alamy Stock Photo:** Björn Wylezich. **25 Dreamstime.com:** Jefunne Gimpel (cra); Elena Moiseeva (crb). **Science Photo Library:** James Bell (br). **26-27 National Geographic Creative:** Carsten Peter / Speleoresearch & Films (c). **30 Dorling Kindersley:** Ruth Jenkinson / RGB Research Limited (bc). **31 Dorling Kindersley:** Ruth Jenkinson / RGB Research Limited (All images). **32-33 Dorling Kindersley:** Ruth Jenkinson / RGB Research Limited (All images). **34 Alamy Stock Photo:** PjrStudio (clb); Björn Wylezich (tr); Science History Images (crb). **35 Dorling Kindersley:** Ruth Jenkinson / RGB Research Limited. **36 Dorling Kindersley:** Natural History Museum, London (bl); Ruth Jenkinson / RGB Research Limited (c, cl). **37 Dorling Kindersley:** Ruth Jenkinson / RGB Research Limited (cra). **Science Photo Library:** Eye of Science (clb). **38 123RF.com:** Konstantin Shaklein (ca); Romolo Tavani (clb). **Dreamstime.com:** Markus Gann / Magann (tr); Vit Kovalcik / Vkovalcik (crb). **Fotolia:** VERSUSstudio (bl). **39 Alamy Stock Photo:** robertharding (b). **Dreamstime.com:** Hotshotsworldwide (cr). **40 Dorling Kindersley:** Ruth Jenkinson / RGB Research Limited (All images). **41 123RF.com:** Dmytro Sukharevskyy / nevodka (c). **Alamy Stock Photo:** Neon Collection by Karin Hildebrand Lau (tr). **Dreamstime.com:** Reinhold Wittich (bl). **iStockphoto.com:** DieterMeyrl (br). **42 iStockphoto.com:** Claudio Ventrella (cl). **42-43 iStockphoto.com:** MKucova (ca). **43 123RF.com:** Kittiphat Inthonprasit (cl). **iStockphoto.com:** ispain (cra); Claudio Ventrella (ca). **Science Photo Library:** Charles D. Winters (br). **44 123RF.com:** Petra Schüller / pixelelfe (cr). **Alamy Stock Photo:** Alvey & Towers Picture Library (c). **Dorling Kindersley:** Ruth Jenkinson / RGB Research Limited (bl). **Science Photo Library:** (bc). **46 Science Photo Library:** Gustoimages (b). **47 Alamy Stock Photo:** Dusan Kostic (bl). **iStockphoto.com:** clubfoto (br). **48-49 Science Photo Library:** Beauty Of Science. **50 123RF.com:** molekuul (cl). **50-51 TurboSquid:** 3d_molier International (b/charred logs). **51 TurboSquid:** 3d_molier International (bc). **53 123RF.com:** mipan (br). **Alamy Stock Photo:** Blaize Pascall (crb). **Getty Images:** Matin Bahadori (clb); Mint Images - Paul Edmondson (bl). **54-55 Benjamin Lappalainen:** blapphoto (c). **56 123RF.com:** Olegsam (bl). **Dorling Kindersley:** © The Board of Trustees of the Armouries (clb/helmet); Natural History Museum, London (clb/Marble). **Dreamstime.com:** Jianghongyan (clb). **Fotolia:** apttone (clb/diamond). **iStockphoto.com:** Believe_In_Me (cra); Belyaevskiy (ca). **57 123RF.com:** Sangsak Aeiddam (bl). **Dreamstime.com:** Nataliya Hora (cl). **Getty Images:** Anadolu Agency (cla); Pallava Bagla (clb); Science & Society Picture Library (clb/Fabric). **58 123RF.com:** bbtreesubmission (bc); yurok (c). **Alamy Stock Photo:** Tim Gainey (clb); Kidsada Manchinda (cra); Monkey Biscuit (crb); Hemis (br). **Dreamstime.com:** Hugoht (bl). **59 123RF.com:** belchonock (cla); Thuansak Srilao (cra); serezniy (cb); sauletas (cr); gresei (bl); Milic Djurovic (bc); Anton Starikov (crb/Jar); Vladimir Nenov / nenovbrothers (br). **Dreamstime.com:** Valentin Armianu / Asterixvs (crb); Dmitry Rukhlenko / F9photos (ca). **60-61 Science Photo Library:** Clouds Hill Imaging Ltd. **62 123RF.com:** Robyn Mackenzie (crb, br); Matt Trommer / Eintracht (cb). **Alamy Stock Photo:** Interfoto (tr); Kristoffer Tripplaar (cl); seen0001 (cb); Anastasiya Zolotnitskaya (bl). **Dorling Kindersley:** Frits Solvang / Norges Bank (cb/Krone). **Getty Images:** © Santiago Urquijo (clb/Bridge). **63 123RF.com:** Manav Lohia / jackmicro (clb/Dime). **Alamy Stock Photo:** money & coins @ ian sanders (clb/Yen); Zoonar GmbH (tc). **Dorling Kindersley:** Gerard Brown / Bicycle Museum Of America (cr). **Getty Images:** David Taylor-Bramley (tr). **iStockphoto.com:** knape (bl). **Photo courtesy Gabriel Vandervort | AncientResource.com:** (clb). **65 naturepl.com:** Alex Hyde (tr). **66 TurboSquid:** Witalk73 (cl). **68 Dreamstime.com:** Jochenschneider (bc). **69 Dorling Kindersley:** The Science Museum, London (ca, cra, cr, c). **Getty Images:** Oxford Science Archive / Print Collector (crb). **72 Science Photo Library:** Tony Mcconnell (tc). **73 Alamy Stock Photo:** Universal Images Group North America LLC (cla). **74 Science Photo Library:** Patrick Landmann (bc). **76-77 TurboSquid:** Witalk73. **78-79 Science Photo Library:** NASA (c). **80 Dreamstime.com:** Markus Gann / Magann (c). **Getty Images:** Digital Vision (tl); Pete Rowbottom (tc); Matthias Kulka / Corbis (crb). **Science Photo Library:** Gustoimages (cl); Edward Kinsman (cr); **81 ESA:** The Planck Collaboration (tl). **ESO:** ALMA (ESO/NAOJ/NRAO), F. Kerschbaum https://creativecommons.org/licenses/by/4.0 (tc). **Getty Images:** William Douglas / EyeEm (bc). **iStockphoto.com:** Turnervisual (cb). **84 Getty Images:** Don Farrall (fcrb); Wulf Voss / EyeEm (c); Melanie Hobson / EyeEm (cr); Francesco Perre / EyeEm (fcr); James Jordan Photography (cb); Steven Puetzer (crb). **85 Science Photo Library:** Andrew Lambert Photography (cl). **88 Alamy Stock Photo:** Alchemy (fcla, cla); Naeblys (bl). **90-91 Juan Carlos Casado:** STARRYEARTH (c). **92 123RF.com:** iarada (bl); Derrick Neill / neilld (cra). **Dreamstime.com:** Aprescindere (bc). **94 123RF.com:** Norasit Kaewsai / norgal (br). **Science Photo Library:** (fcrb, fbr); Tek Image (tr); Martyn F. Chillmaid (crb). **97 Alamy Stock Photo:** geogphotos (br). **102 Dreamstime.com:** Antartis (crb). **103 Dreamstime.com:** Markus Gann / Magann (tl). **106-107 TurboSquid:** Zerg_Lurker. **107 iStockphoto.com:** Mikita_Kavalenkau (cb). **112-113 NASA:** WMAP Science Team (tr). **112 NASA:** NASA / ESA / S. Beckwith(STScI) and The HUDF Team (cb). **Science Photo Library:** Take 27 Ltd (clb). **113 NASA:** WMAP Science Team (crb). **114-115 Science Photo Library:** Mark Garlick. **114 NASA:** JPL-Caltech / ESA / CXC / STScI (cr). **Science Photo Library:** David A. Hardy, Futures: 50 Years In Space (bl). **115 Dreamstime.com:** Tose (br). **Getty Images:** Robert Gendler / Visuals Unlimited, Inc. (ca). **iStockphoto.com:** plefevre (cla). **NASA:** ESA / JPL-Caltech / STScI (cra); JPL-Caltech (clb); X-ray: NASA / CXC / SAO / J.DePasquale; IR: NASA / JPL-Caltech; Optical: NASA / STScI (cb); ESA, S. Beckwith (STScI) and the Hubble Heritage Team (STScI / AURA) (crb). **118-119 National Geographic Creative:** NASA / ESA (c). **120-121 Science Photo Library:** NASA. **120 Alamy Stock Photo:** Science Photo Library (clb). **Dreamstime.com:** Torian Dixon / Mrincredible (cla). **121 Getty Images:** Photodisc / StockTrek (tr). **123 Dreamstime.com:** Gregsi (tc, cra). **125 Alamy Stock Photo:** TAO Images Limited (tr). **127 Alamy Stock Photo:** Science History Images (br). **129 123RF.com:** mihtiander (crb). **Getty Images:** Sirachai Arunrugstichai (fcrb). **133 Alamy Stock Photo:** World History Archive (bl); Z4 Collection (cb). **Dorling Kindersley:** The Science Museum, London (crb). **Dreamstime.com:** Anetlanda (tc); Koolander (cla); Bolygomaki (clb). **134 Science Photo Library:** Eye of Science (cl). **135 123RF.com:** Eduardo Rivero / edurivero (br). **Dreamstime.com:** Andrey Sukhachev / Nchuprin (cra/Bacteria); Peter Wollinga (crb). **Getty Images:** Roland Birke (cr/Protozoa). **Science Photo Library:** Dennis Kunkel Microscopy (cra); Power And Syred (bc); Gerd Guenther (cr). **136 Science Photo Library:** Chris Hellier (cb). **137 Alamy Stock Photo:** Mopic (cra). **Dreamstime.com:** Steve Byland / Stevebyland (br). **Gyik Toma / Paleobear:** (tl). **138 Alamy Stock Photo:** Dave Watts (tr). **139 123RF.com:** Iakov Filimonov / jackf (tr); Sergey Krasnoshchokov / most66 (cra); Christian Musat (crb/Spectacled bear); Pablo Hidalgo (bc). **Dreamstime.com:** Mikhail Blajenov / Starper (crb); Guoqiang Xue (cr); Ivanka Blazkova / Ivanka80 (cr/Sun bear); Minyun Zhou / Minyun9260 (br). **140 Science Photo Library:** Steve Gschmeissner (bl). **141 Science Photo Library:** Eye of Science (crb). **142 Science Photo Library:** Steve Gschmeissner (cl). **146 Science Photo Library:** Dr Jeremy Burgess (cb). **147 Science Photo Library:** Dennis Kunkel Microscopy (br). **149 Getty Images:** wallacefsk (cr). **iStockphoto.com:** BeholdingEye (fcr). **150 Dorling Kindersley:** Jerry Young (ca). **159 123RF.com:** Anastasija Popova / virgonira (br). **161 Alamy Stock Photo:** FLPA (cb). **Getty Images:** Kiatanan Sugsompian (bc). **162 June Jacobsen:** (cl). **166 Dreamstime.com:** Haveseen (tr); Worldfoto (tl). **Getty Images:** Visuals Unlimited, Inc. / Ken Catania (c). **iStockphoto.com:** lauriek (tc). **167 Alamy Stock Photo:** blickwinkel (cr). **Dorling Kindersley:** Jerry Young (crb/Monkey). **Getty Images:** De Agostini Picture Library (cra/Bat); Yva Momatiuk & John Eastcott / Minden Pictures (tl); Nicole Duplaix / National Geographic (cra). **iStockphoto.com:** arlindo71 (br); sharply_done (crb). **168 iStockphoto.com:** GlobalP (br). **Science Photo Library:** Omikron (tl). **169 Dreamstime.com:** John Anderson / Johnandersonphoto (cra). **172 Dreamstime.com:** Stu Porter / Stuporter (br). **iStockphoto.com:** TommyIX (cl). **173 Getty Images:** Gail Shumway (t); Alexander Safonov (cb). **174 Alamy Stock Photo:** garfotos (tr); Shoot Froot (cra); Richard Garvey-Williams (cl); John Richmond (crb); Brian Haslam (br). **Depositphotos Inc:** danakow (cr). **175 Alamy Stock Photo:** Brian Haslam (cla). **Harald Simon Dahl:** www.flickr.com/photos/haraldhobbit/14007088580/in/photostream (tc). **Getty Images:** Alan Murphy / BIA / Minden Pictures (bl). **SuperStock:** Konrad Wothe / Minden Pictures (cra). **177 Alamy Stock Photo:** Premaphotos (cb); Poelzer Wolfgang (tl). **Getty Images:** David Doubilet (clb); Brook Peterson / Stocktrek Images (tc, tr); Stephan Naumann / EyeEm (crb). **183 Getty Images:** Tim Laman / National Geographic (c); John E Marriott (bc). **184 Alamy Stock Photo:** age fotostock (br). **Dreamstime.com:** Chase Dekker (cra/Taiga); Max5128 (cra); Snehitdesign (c); Jeffrey Holcombe (crb). **185 Dreamstime.com:** Eddydegroot (bl); Denis Polyakov (cla); Ivan Kmit (cra); Szefei (crb); Zlikovec (br). **iStockphoto.com:** ianwool (tl). **186-187 Depositphotos Inc:** Olivier26. **187 iStockphoto.com:** pamspix (br). **188 Getty Images:** Bill Curtsinger / National Geographic (cla). **190 123RF.com:** Anan Kaewkhammul / anankkml (ca). **Alamy Stock Photo:** Mark Daffey (fbl). **Dorling Kindersley:** Cotswold Wildlife Park (tr). **Dreamstime.com:** Natalya Aksenova (clb); Johan Larson / Jaykayl (bl); Wrangel (cl, br); David Spates (c); Anton Ignatenco / Dionisvera (bc); Sailorr (cr). **191 Alamy Stock Photo:** dpa picture alliance (cb). **Dorling Kindersley:** Twan Leenders (cra/Turtle). **Dreamstime.com:** Frozentime (cra/Kingfisher); Isselee (cl); Meunierd (crb/Deer). **194 Science Photo Library:** Sinclair Stammers (tr). **195 Dreamstime.com:** Koolander (cla); Daniel Prudek (tl). **NASA:** JPL-Caltech (bc). **196 Dorling Kindersley:** Rotring UK Ltd (tr). **Dreamstime.com:** Dave Bredeson / Cammeraydave (crb); Tanyashir (bc). **iStockphoto.com:** artisteer (cb). **197 Alamy Stock Photo:** Tetra Images (cr). **Dreamstime.com:** Yu Lan / Yula (c). **iStockphoto.com:** Icsatlos (crb); seb_ra (cb). **198 123RF.com:** Koji Hirano / kojihirano (fcrb); Eric Isselee / isselee (cr/Gibbon). **Dorling Kindersley:** Andrew Beckett (Illustration Ltd) (cb, crb); David J Patterson (ftr); Jerry Young (cl, bl). **Dreamstime.com:** Isselee (clb); Andrey Sukhachev / Nchuprin (tr). **199 123RF.com:** Andrejs Pidjass / NejroN (cra). **Dorling Kindersley:** Natural History Museum, London (tc); Jerry Young (clb, crb). **Dreamstime.com:** Isselee (fcra); Janpietruszka (cla); Piotr Marcinski / B-d-s (br); Volodymyrkrasyuk (cl)

All other images © Dorling Kindersley
For further information see: **www.dkimages.com**